Streaming Devices + Streaming Services

Reviews, comparisons, and step-by-step instructions

I0464496

By Ken Wickham

Section 107 contains a list of the various purposes for which the reproduction of a particular work may be considered fair, such as criticism, comment, news reporting, teaching, scholarship, and research. Section 107 also sets out four factors to be considered in determining whether or not a particular use is fair.

kwickham.kw@gmail.com

Published in the United States of America

Wickham, Ken.
Streaming Devices + Streaming Services/ Ken Wickham.

First Edition : June 10, 2014

Second Edition: April 20, 2014

ISBN-13: 978-1511833684

14 13 12 11 10 / 10 9 8 7 6 5 4 3 2 1

Dedication

This Kindle book is dedicated to all the antenna suppliers, video streamer devices, and free online media content providers

Thank You.

Ken

Forthcoming in this series

Internet + Wi-Fi phones

Also Available

Volume 1

Antennas + TV Program Guides
ISBN-13: 978-1499321135
ISBN-10: 1499321139
Kindle ASIN: B00K1M63SK

Or combined Vol 1 + 2

Antennas & Streaming
SBN-13: 978-1500399986
ISBN-10: 1500399981
Kindle ASIN: B00KWPCS9Y

Table of Contents

Review and Instructional text copyright © 2014, 2015 Ken Wickham 2

All Rights Reserved 2

Dedication 3

Sick of expensive unused channels 10

How many of the paid channels do you use? 10

How many TV shows come from network broadcast television? 10

Rise of the Cord Cutters 10

Overview of the book 11

Planning your TV entertainment services 13

I. Streaming Device and Streaming Service 13

Streaming Internet Speed minimum recommendations 13

II. What TV shows, series, and movies do you watch? 14

Which internet sports packages do you want or already have? 14

What Kind of Viewing TV or Computer Tuner do you own? 17

TV Connectors 17

TV resolution, aspect ratio, formats 18

Modern signal types 18

The i and p in the resolution 19

Old analog signals (for comparison) 19

Current HDTV signals 19

Content Search Tool Website 20

1.) TV.com 20

1A.) TV SHOWS/EPISODES 20

1B.) MOVIES 21

2.) CanIStream.It 22

2A.) TV Shows Episodes 23

2B.) MOVIES 25

3.) Watchily 25

4.) eTRIZZLE.com [hasn't been updated since 2013] 26

4A.) TV Shows and Episodes 26

4B.) MOVIES 28

Income base of streaming services 30

Commercial Based 30

Subscription Based 30

On Demand, Pay-per-viewing, Pay-as-you-go 30

My recommendations for Streaming Device Choices 30

Best device USB dongles: 31

Streaming Devices 32

Streaming Devices you may already own 32

Computer, tablet, smartphone, laptop 32

Change TV mode 33

Video Game Consoles 35

Xbox 35

Wii 36

Playstation – PS3, PS4, PS VITA 38

Smart TV 40

Blu-ray Player 40

Buy a Streaming Device 41

Chromecast 41

Amazon Fire TV & Fire Stick 44

G-Box Q & Midnight MX2 XBMC 45

Roku 1, 2, 3, LT, Streaming stick 47

WD TV Live 48

Apple TV 49

NETGEAR Neo TV, Neo Max 51

D-Link MovieNite & Movie Nite Plus 52

RCA Wi-Fi Streaming Media Player 53

VIZIO Co-Star and Co-Star LT 54

Android Mini PCs 55

Streaming Apps for Mobile and Mini Devices 58

Android Apps 58

Streaming Sources 61

My Recommendations for streaming services 61

Streaming Service Providers 62

Hulu (NBC, ABC, Fox, the CW, ION, WB, BBC, MTV, BET, VH1, Nat Geo, Syfy, Spike, TBS, TNT, USA, Oxygen, Lifetime, LMN, E!, AMC, WWE, Food Network, Disney, Comedy Central, A&E, Bio, DIY, HGTV, History, H2, TVLand, Travel Chan, IFC, Jim Henson, Cartoon Network, Adult Swim, Nickelodeon, PBS Kids, Anchor Bay, Image Ent, Lions Gate, MGM, Anime Net, Aniplex, Funimation, Bandai, Manga, TokyoPop, TOEI, VIZ, Exercise TV, GAIAM TV, Fora TV 62

Movie and TV Library 62

Hulu Original Series 63

Popular TV Shows 63

Popular Movies 64

CBS All Access 65

Netflix (EPIX) (Back catalog Time Warner, Universal, MGM, Paramount, Lions Gate, Sony, 20th Cent Fox, Disney) 66

Original Content 66

Netflix Instant Video 67

Popular TV shows 68

Completed Series 68

Amazon Instant Video (EPIX) 68

Amazon Instant Video Membership 68

Movie and Video Content 68

Amazon Prime Popular Movies 69

Popular TV shows 69

Original Content 70

Sling TV 70

Playstation Vue 70

Redbox Instant (EPIX) - Shutdown Oct 10, 2014 71

See message of shutdown http://about.redboxinstant.com/news/ 71

Redbox Kiosk available titles. http://www.redbox.com/ also has links to iOS and Android apps 71

HBO Go 72

HBO NOW 72

Crackle (Sony) 72

   Movie and TV Library 72

   Content Partners 72

   Genres 72

   Top Titles 73

   Movies 73

   TVShows 73

YouTube Movies (Google: Merged with Google Video) 74

   Movies by 74

Popcornflix (Screen Media) 74

   Top Movies 74

   TV Series 75

   National Geographic Channel 75

PlutoTV 75

TubiTV 75

Snag Films 75

Filmon 76

VUDU (Walmart) 76

Target Ticket 76

M-GO (Dreamworks and Technicolor) 76

Sony Entertainment Network 77

Fandor 77

ErosNow 77

DramaFever 77

Sports Packages   78

    MLB   78

    MLS   78

    NASCAR   78

    WNBA   78

    NFL (Computers and Windows, Android, iOS Tablets)   78

    NHL   79

    NBA   79

    UFC   79

    WWE Network   79

        Popular Free Sports Online and Apps   79

NETWORK TELEVISION (Online Website Links and Resources)   80

    Commercial Television   80

    Public Telvision   80

    Childrens   80

    Movies   80

    Music   80

    Living   81

    Classic TV   81

    Religious   81

Paid TV channels (Online Website Links and Resources)   82

    NOTES   86

Sick of expensive unused channels

$157 monthly. $1884 annually.

That's how much money we were paying for cable television and internet. Sure we had 300+ channels. However, how many did we actually use?

How many of the paid channels do you use?

We watched two singing contest shows, a few suspense thrillers, a comedy sitcom, and a couple of sci-fi series. We actually only watched four TV shows at any given time, due to different show seasons. Occasionally, we rented a movie from the big red colored box outside store, or ordered through the mail some rental movie internet videos from a monthly subscription. We did this because we didn't want to wait a year to watch these just-out-of-the-theater movies.

I noticed quickly, that the movies we received from the premium channels, we had already seen through the box, received via the mail or online streaming. There was only one premium TV show series that we watched 3 months every year.

How many TV shows come from network broadcast television?

Most of the TV shows we watched came from network broadcast television. Only three cable television series came from the part of cable that we paid. The portion that we paid for premium cable was about $100, only to watch three shows, and an occasional interesting history or biography documentary. That is $1,200 a year for three shows plus a little here and there.

Rise of the Cord Cutters

Sick of paying all this money for almost nothing, I had to think of an alternative. Many have cut their cable bills entirely off and become what many call "cord cutters".

Companies that have an eye towards the future are heading this wave of personalized-on-demand, subscription, or free commercial paid content.

Netflix, Google, Hulu, Amazon and even a few cable and satellite such as Xfinity, Dish Network, and DirectTV are companies are heading towards mobile, and more personalized content. TV Broadcasters are filling their websites with streamable episodes and content. A few live streams of news and a few networks already are on the net.

TV media devices are quickly becoming a part of everyday life.

Overview of the book

This book has been to help you know what streaming services and devices are available. The book will begin with an overall checklist of major steps to accomplish setting up services. Next, you will get to know the type of connectors, tuner, and resolution of your current TV or computer that you will be using to receive streaming services from the internet. You will figure out what content you currently own, keeping in mind the possibility of integrating that content into the streaming system. You will write down what content you desire. You will then figure out what formats, providers, and devices that content may be available, through using mainly four free online internet search services. My device recommendations for different situations will be presented. Then brief information will be presented for major streaming technology and devices available currently. With this information, you may begin to figure out your own recommendation if you would rather figure it out on your own, rather than following my suggestions. A few feature summary tables will be presented, as well as summary of device prices, contents, hardware, and setup steps for each of the devices. I give my overall steps to selecting streaming services in an attempt to replicate pay TV to an extent. I then give my streaming service recommendations for various situations. Streaming services will then be introduced with a brief summary of available content, prices, and popular content. After that, sports packages will be presented with the associated costs and brief major features. Finally, lists of major broadcast TV and Pay TV online internet sites with links/addresses will be presented with a small summary of available content such as streaming live content, streaming episodes, and clips.

The book intends to help guide you in setting up your own internet streaming services. It does not cover older model streaming devices, however there may be some similarity with manufacturer newer models. The goal is to replace your pay TV massive bill with smaller fees, more customized content, and near free yet legal.

Steps to receiving streaming service

1. [ ] First write down the qualities of your TV or computer monitor you will be watching. This information includes connector types, TV tuner, and screen resolution. Also record your internet service speed. Record this info on the planning information sheet in section I. You may print off a copy of the planning pages.

2. [ ] If desired, fill out the movie inventory sheet(s) listing all of your current movies. I provide one that you can use by printing off as many needed copies. On that sheet, I make available room to record other information that may help you keep track of your movie library. This may help you in deciding how to integrate existing content into your streaming services and overall entertainment system.

3. [ ] Decide what type of streaming content you wish to see. You can do section 2 filling out your favorite content, and using the four free online search website tools to fill out most of the content supplier availability information. See the instructions, for quicker help gathering where these TV shows and movies are located.

4. [ ] If you will base your content on a device you already own, what is the content limited to that device? Use the charts at the beginning of the streaming device section to compare content available on popular devices. If you will be purchasing a device, make sure to consider the limited content available to the devices during your decision process. Make sure you can receive and use the content which you wish to use. If you have no priority of content, realize some content you might not be able to receive on different devices. Also consider the speed of the device and internet connection.

5. [ ] If needed, purchase your device either based on price, desired content, features, speed, TV connections, and content reception options.

6. [ ] Set the device up to the internet service.

7. [ ] Set up the content you already possess.

8. [ ] Purchase, subscribe, or configure the streaming services that you will be using. Subscriptions and services may offer some free time or number for downloading. I make my recommendations in the Streaming Service Section.

With this overview of the steps needed to plan, analyze, and set up your streaming services, we will first look more at a series of pages you can print off, with this checklist, to help you do steps 1 through 3.

For a printable PDF of these steps and charts, I make them available off of my Google Drive. Click on the link or direct your browser to this address.

Printable PDF https://drive.google.com/file/d/0B2I_GEXbFycdOE4zWG9SLUYyMm8/view?usp=sharing

Planning your TV entertainment services

I. Streaming Device and Streaming Service

[Print these pages off if needed] See the chapter "What Kind of Viewing TV or Computer Tuner do you own?"

In the book or on separate pieces of paper, please write down the following information in order to help you setup an alternative to expensive paid TV

What type of television connectors does your TV have? How many?

- Coaxial _____
- HDMI _____
- S-Video _____
- Composite (LR(red white) audio, (yellow) video _____
- Component (Green - Y, blue – Pb, red – Pr; all three are video, audio is separate)
- USB _____
- VGA _____
- DVI _____

What kind of TV tuner will you be using? (Check or write down each one)
  A. Built in TV [ ]
  B. TV converter box [ ]
  C. Computer TV tuner [ ]

What is your TV resolution?
  A. 480i (analog) [ ]
  B. 480p (digital) [ ]
  C. 720p [ ]
  D. 1080p [ ]
  E. WQHD 1440p [ ](Highest currently possible)

What is your Internet download speed? Search Google "*ISP name* speed test", filling in your ISP provider.

_____Mb/s

Streaming Internet Speed minimum recommendations

(Circle your internet speed range)

Laptop 1 Mb/s
SD TV 1-2 Mb/s
HD TV 720p 2-4 Mb/s
HD TV 1080p 5-9 Mb/s

II. What TV shows, series, and movies do you watch?
www.tv.com, http://www.canistream.it/ , www.eTRIZZLE.com

1.) Fill out the first form, the list of content on the sheet on the following page, which you currently watch or plan on watching. Start with sports packages if desired, then movie/TV shows.

Which internet sports packages do you want or already have?

NBA	MLB	NHL	MLS	NFL	UFC	MMA	WWE	NASCAR	WNBA

Next you can make a list of all the content that you already own. I have created an inventory list that you can use to record this and other important information. Some of these providers only provide live audio some provide live internet. Check the sports section to see which are live, and which are next day.

- If you have a numbering system, you can number your movie or show content in the first column

Movie – You can put the name of the movie or show in the 2nd column. If you are organized you can use any form of organization such as alphabetical, by genre, by acter/actress, etc…

Part in Series – I created this column for movies or shows in some sort of series. It can be a movie series or TV show season/episodes. You might want to indicate in this column something such as "Season 6" or episodes 1 to 4, or part 2 in a trilogy.

Genre – Here you can put categories such as Sci-Fi, Comedy, Romance, Action, Adventure, Fantasy, Anime, Horror, or Documentary to give you a few examples.

Minutes – Length of the movie, episode, show, DVD, or whatever.

Rating – This can either be something like PG, R, PG-13 or better yet your own rating system such as 1 to 5 star or A through F. It can be something simple like good, great, sucks, hated.

Format – Is this ownership content in the form of which of these categories

Location – Where is this content located? Insurance may want a list if your movie collection is destroyed in some disaster or fire/flood/tornado.

Value/Cost/Price – Insurance may want a list of the value of your movie/show collection if perishable.

2.) Next, fill out the 2nd form of content desired, listing shows, series, and movies that you wish to see. Use the three *Content Search Tool Website.* See that section for details and instructions.

www.tv.com, http://www.canistream.it/ , www.eTRIZZLE.com

#	Movie	Part in serie	Genre	Minutes	Rating	Format	Main Actors/Actresses	Value/Cost/Price	Location
						○ Blu-ray ○ DVD ○ Digital ○ VHS			
						○ Blu-ray ○ DVD ○ Digital ○ VHS			
						○ Blu-ray ○ DVD ○ Digital ○ VHS			
						○ Blu-ray ○ DVD ○ Digital ○ VHS			
						○ Blu-ray ○ DVD ○ Digital ○ VHS			
						○ Blu-ray ○ DVD ○ Digital ○ VHS			
						○ Blu-ray ○ DVD ○ Digital ○ VHS			
						○ Blu-ray ○ DVD ○ Digital ○ VHS			
						○ Blu-ray ○ DVD ○ Digital ○ VHS			
						○ Blu-ray ○ DVD ○ Digital ○ VHS			

TV Show or Movie	Website	Netflix	Hulu	Hulu Plus	Amazon Prime	Amazon Instant	HBO GO	VUDU	iTunes	GooglePlay	Crackle	YouTube	Popcorn Flix	MGO	CinamaNow	PLEX	EPIX

What Kind of Viewing TV or Computer Tuner do you own?

The first step you need to do is to figure out if your TV has a built in digital TV tuner, or is an older analog TV tuner. If you are going to use a computer as a TV, you will need to check if you have a TV tuner card (normally indicated by a coaxial input port on your computer).

Built into TV [ ]
Converter Box [ ]
Tuner Card [ ]

At the same time, you may want to make a list of all the devices you plan on including in your entertainment system.

Accessories
Antenna [ ]
Media Player [ ]
Media Streamer [ ]
DVR [ ]
DVD player [ ]
Blue-ray player [ ]
Video game console [ ]
Sound system or speakers [ ]

TV Connectors

Next, you need to know what type of connectors are included on your television(s) you will be using.

Return to Table of Contents

HDMI

HDMI emerging in 2004, HDMI has replaced coaxial as the high definition connection of modern televisions. 90% of HDTVs by 2007 had HDMI connectors. By 2009, all digital televisions had at least one HDMI connectors. Many converters exist to convert different type of connectors and wires into HDMI. The signal probably will not play at the highest resolution however. Be cautious of having too many connectors, splitters, and wires which leads to signal loss and overall masses of wire octopuses.

S-Video

S video has a max 480i/576i signal definition. S-Video is slightly better than composite video, using 2 channel encryption instead of one. You may have to use this for older TVs though composite a/v is normally more common.

Composite A/V

Older TVs might need to use the composite port. Composite has a max of 480i/576i. This is the 3 color wires of yellow for video, and red and white for audio. Composite A/V is slightly lower quality compared to S-Video because it only uses one channel instead of the two that S-Video uses.

Ethernet

Cat-5e is the current standard for Ethernet wiring. Ethernet wiring is mainly used to connect devices by wire to routers and modems.

Coaxial

Older still is the basic coaxial input and output. RG-6 is the current standard for coaxial cable.

USB

USB port on TVs can run movies, listen to music, and look at pictures from a thumb drives. It can also update the TVs firmware by downloading the software putting it on a thumb drive then inserting the drive into the port. External hard drives can also be used to play content connecting directly to the TV.

TV resolution, aspect ratio, formats

When the entire United States switched over from analog to digital service, the entire country was thrown into a little frenzy. What emerged however has its advantages and disadvantages.

Modern signal types

Broadcaster digital terrestrial television (DTV) broadcasting. Broadcasters normally transmit one or both types of picture formats, which vary in size and aspect-ratio. High definition television (HDTV) for the transmission of high-definition video and standard-definition television (SDTV).

The i and p in the resolution

In the signal transmission the gap is called *interlacing video* (symbol = i) in the analog method and *progressive* (symbol = p) for digital video. In interlacing they sent half a frame at a time, first the odd lines then the even lines. Now they send line-by-line. So when you read 720p, this actually means 1280x720 digital *progressive scan* signal in that it receives a signal line by line. This reduced eyestrain from interline twitter making images and movement more smooth.

Old analog signals (for comparison)

CGA computer monitors had a resolution of 320x200 up to 640x200 4 bit 16 color.

VGA computer monitors (1987) introduced 640x480 up to 800x480 16 bit in 4:3 aspect ratio. VGA has 256 colors in 320x200 mode. The second resolution is the VHS and Beta-max resolution. 480i is also the broadcast resolution of old analog televisions.

Current HDTV signals

HD 480p has 640x480 pixel or 4:3 aspect ratio (4 units wide by 3 units high) is SDTV. This is closest to DVD quality which is 720x480 which can be shrunk to fit SDTV, or play in wide-screen mode with dark space above and below the picture.

HD 720p widens the resolution to 1280x720 16:9 aspect ratio.

HD 1080p resolution of 1920 × 1080 at a 60 Hz with 16:9 aspect ratio. This is Blue-ray resolution.

With a few pages of content plans, a little knowledge about TV connections, resolution, and wire connections, we can now look into the basics of streaming. We can then look at some of the popular streaming devices. We can compare device features, characteristics, advantages, and disadvantages, and prices.

What is streaming video? And what are the main basis of streaming services?

Return to Table of Contents

Content Search Tool Website

www.tv.com, http://www.canistream.it/ , www.watch.ily, www.eTRIZZLE.com **Tools to search multiple online content providers.**

COMPARISON OF CONTENT SEARCH TOOLS

Search	Hulu Plus	Netflix	Amazon Instant	Amazon Prime	Redbox Kiosk	GooglePlay	YouTube	Crackle	Popcornflix	VUDU	MGO	iTunes
TV.com	√	√	√	√		√				√		√
CanIStream.It	√	√	√	√	√	√	√	√		√		√
Watchily		√	√									√
etrizzle.com	√	√	√	√	√	√		√		√	√	√

Search	CinamaNo	PLEX	EPIX	XBox	Sony Ent	Target Ticket	SnagFilms	Xfinity St	HBO GO	Showtim	Cinama	Fandor
TV.com								√	√			
CanIStream.It			√	√	√	√	√	√	√	√	√	√
Watchily								√	√	√	√	
etrizzle.com				√	√	√			√			

This table is shown to show you which search website will search which provider services. Notice that none of these three searches CinamaNow, PLEX, or Popcornflix. Use these to fill in the second sheet of desired content. I show an example for searching for a TV show, in this case HBO's Game of Thrones, and a movie, in this case Resident Evil.

1.) TV.com

1A.) TV SHOWS/EPISODES

To help you fill out the rest, go to www.tv.com, 1) click in the *Search TV.com* box and type the title. Find the show or movie logo or picture, then look for and click the **Watch** link near the bottom of that section.

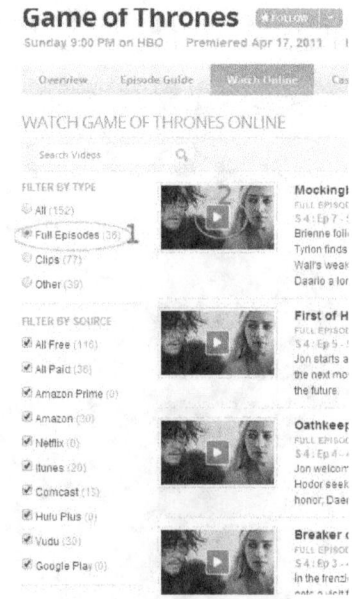

1.) Click on the Full Episodes filter for TV shows. 2) Hover over play button for play options (hover shows results below). Options are FREE, SUBSCRIPTION, or BUY IT for TV show episodes.

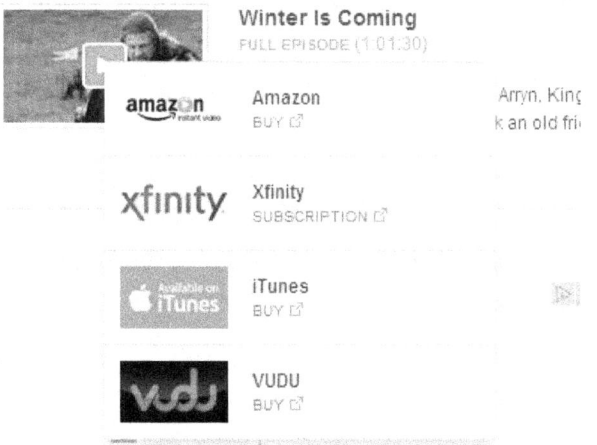

1B.) MOVIES

For movies, 1.) you can click on the movie in the result section.

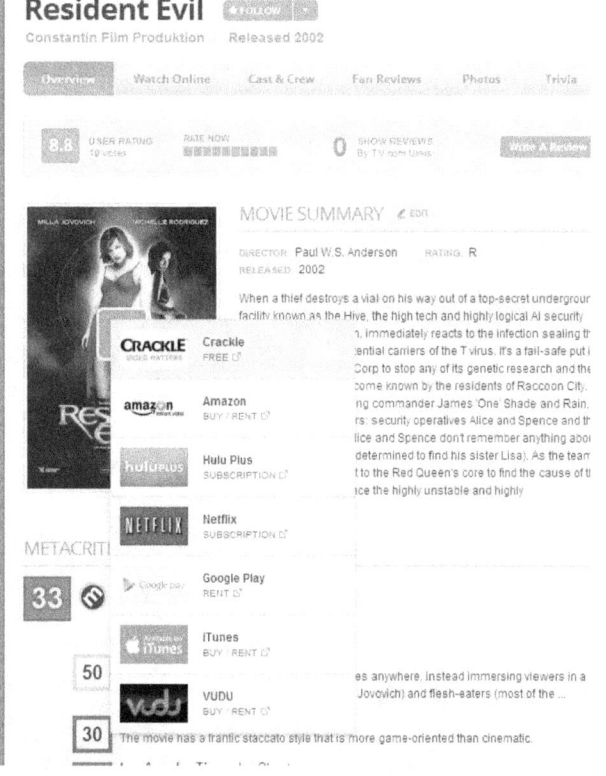

2.) Then hover over the **play** button. Although this does not give ALL the available options, it is a start. It gives the content for Amazon Prime, Amazon Instant, Netflix, iTunes, Hulu Plus, VUDU, and Google Play which are the big guns of streaming services. It will note if the method is FREE, BUY, RENT, or SUBSCRIPTION based from each of the providers.

2.) CanIStream.It

Next, you can also use http://www.canistream.it/ which searches four main areas usable by cord cutters: streaming, digital renting, digital purchase, and disc purchase/rentals.

For **streaming,** it searches Netflix, Amazon, Hulu Plus, as well as Crackle, YouTube, EPIX, Xfinity Streampix, Fandor, and Snag films.

For **digital renting,** it searches Amazon, iTunes, Google Play, VUDU, YouTube, and Sony Entertainment Network, Target Ticket.

For **digital purchasing**, it searches Amazon, iTunes, Google Play, VUDU, Xbox 360, Sony Entertainment Network, Target Ticket.

For **disc Purchase and Rental**, it searches Amazon (both DVD and Blue-ray), Netflix, and Redbox.

Also Xfinity subscribers can check packaged channels through CanIStream.It

Return to Table of Contents

2A.) TV Shows Episodes

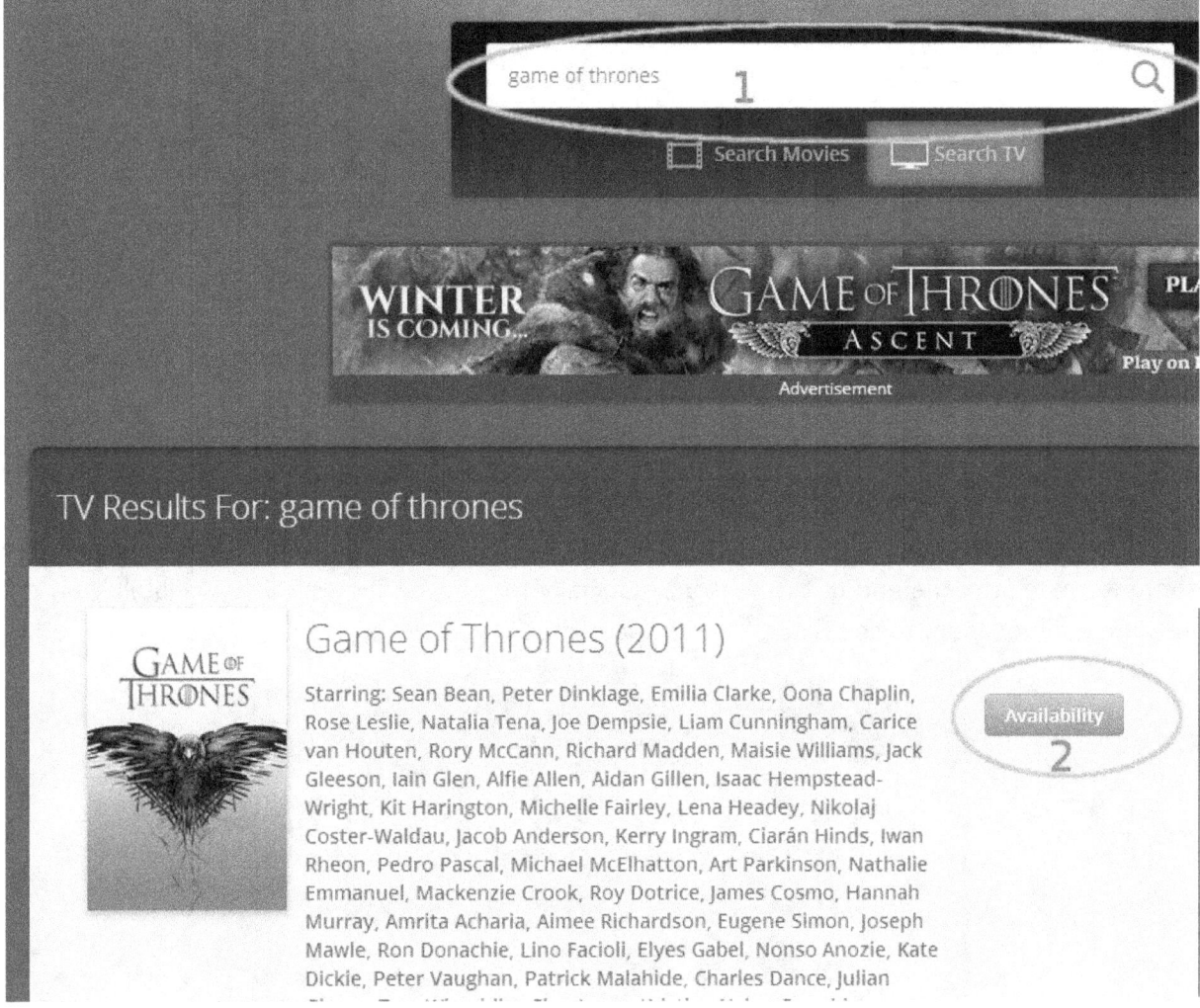

1.) First enter the title in the search box. Next, be sure to indicate underneath whether it is a **Movie** or a **TV** program. Click on TV for this search.

2.) Click on **Availability**.

Return to Table of Contents

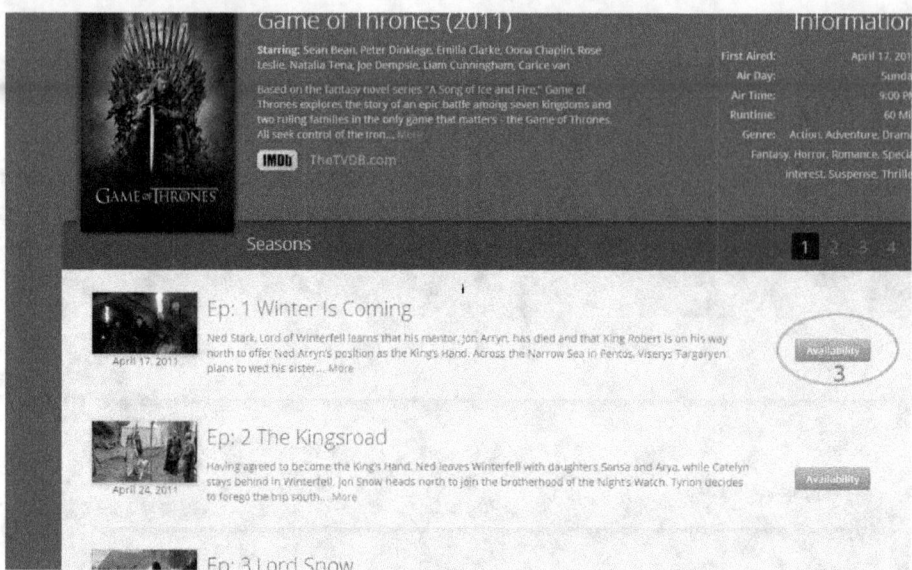

3.) Look for the episode that you are wanting to watch. Some episodes will not be available on some formats until the end of the season, especially for pay TV shows. Click Availability on the episode that you want to see.

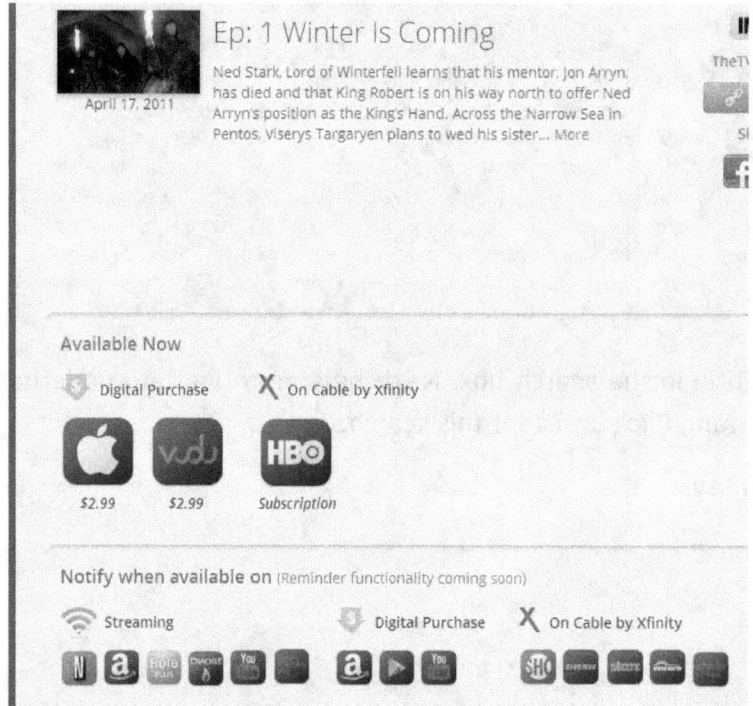

4.) The icon will show Streaming prices, ads for free streaming, or subscription and digital purchase prices. At this moment you cannot rent episodes. Also you can select Notify When available on select services, if signed in.

3.) Click on the service icon to go to that service.

2B.) MOVIES

For movies the search is quicker.

1.) Type the title of the movie in the search box. Make sure Search Movies is selected below the search box.

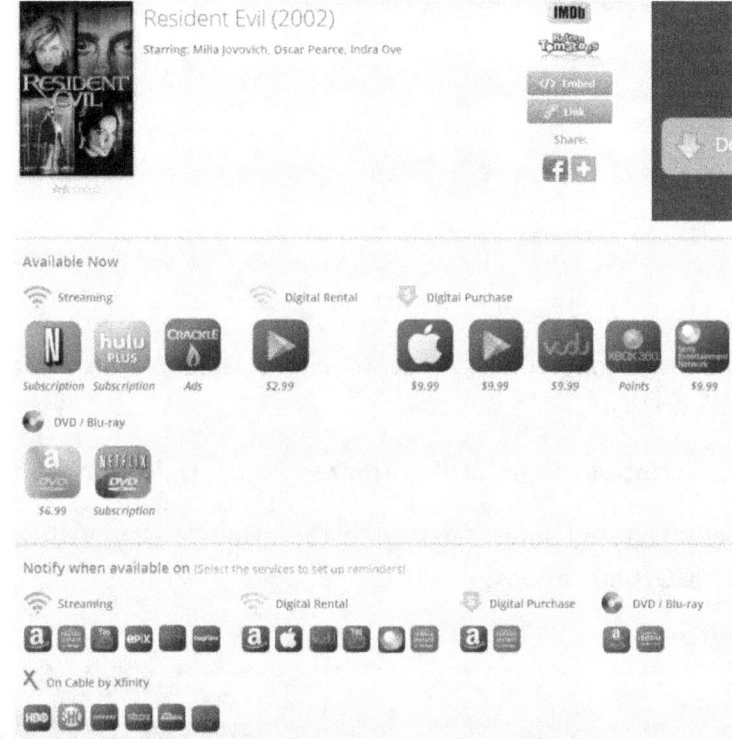

2.) It will return the results listing streaming options. Digital rental, digital purchase, and DVD/Blu-Ray. Also you can select Notify When available on select services, if signed in.

Prices will be shown for purchasing/rental. Free will mention Ads, Xbox uses points. Subscription services will be mentioned below the symbol.

3.) Click on the service icon to go to that service.

Return to Table of Contents

3.) Watchily

http://watchi.ly/

This will search for content according to some of what is available for your device. You just enter the name of the show or movie, select your device. Then click **Go**.

It searches Amazon Instant, Netflix, HBOGO, iTunes, MAXGO, Showtime Anytime, XFINITY

Devices supported: Roku, Amazon Fire TV, Apple TV, Chromecast, Google TV, iPhone, Android, Smart TVs, Video game consoles - PS, Xbox, Wii

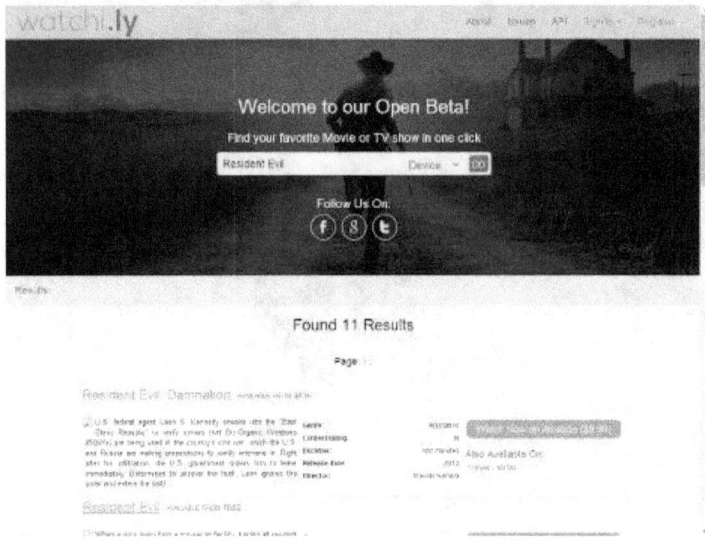

4.) eTRIZZLE.com [hasn't been updated since 2013]

eTrizzle searches TV shows from iTunes, Amazon Instant, Hulu, Netflix, Amazon Prime, Crackle, HBO Go, Google Play, and VUDU.

Movies will search **subscriptions** from Netflix, Amazon Prime, Hulu, and Crackle.

Movies will search **rentals** from iTunes, Amazon Instant, Google Play, VUDU, Target Ticket, Xbox video, Sony Movies, MGO, and Redbox.

4A.) TV Shows and Episodes

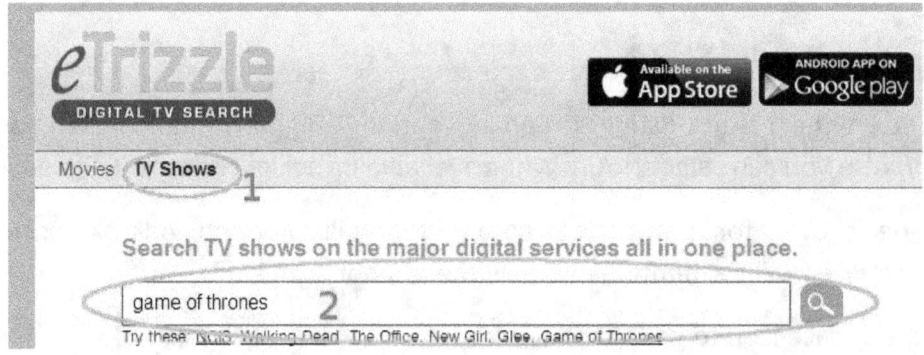

1.) Select TV Shows.

2.) Type in the search window the name of the TV show or episodes you are looking for.

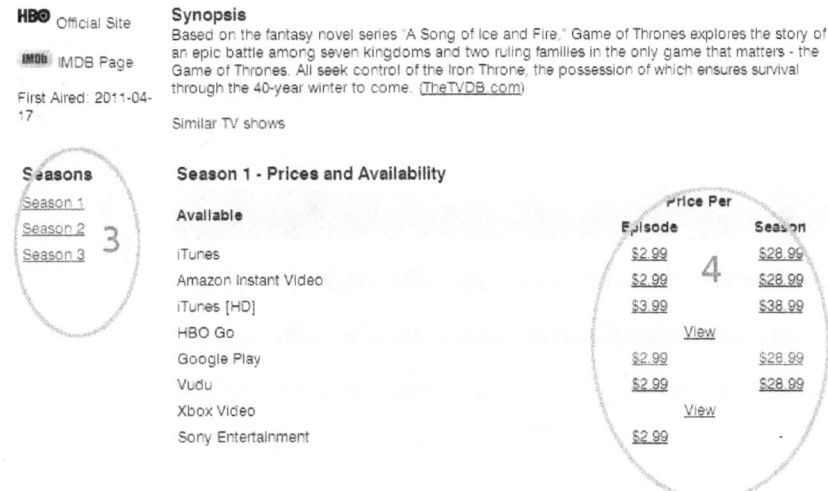

3.) Although the latest season is listed on top, scroll down to see past seasons. Click on which season you want.

4.) Available subscription, free, and purchase of episodes will be indicated. Click on the desired service indicate to the right of the service provider.

Return to Table of Contents

4B.) MOVIES

1.) Make sure **Movies** is selected.

2.) Enter the title of what movie you will look for in the search box. Press enter or click the search icon.

3.) If multiple titles have the same name, click on which movie out of the list is the one that you are trying to find.

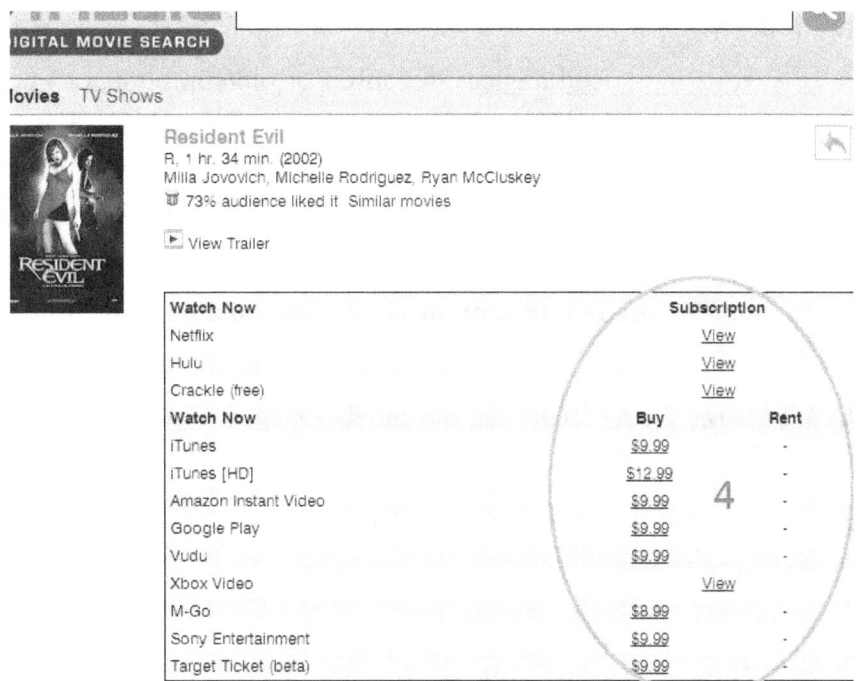

4.) Look at the available services. Select which one you prefer.

Return to Table of Contents

Streaming Video

Streaming video is something anyone can do with a high speed internet connection. There are three main basis of streaming services: commercial, subscription, and on pay-as-you-go.

Income base of streaming services

Commercial Based

Some networks offer commercial based on demand episodes and movies. These commercial based services ingrain multiple commercials per viewing.

Subscription Based

Paid subscription based services charge a fee per month. The paid subscription services offer a catalog of available movies which change over time.

On Demand, Pay-per-viewing, Pay-as-you-go

Instant video allows you to rent and buy video episodes and movies on demand.

With a basic knowledge of the three basis for streaming services, we can focus more on the devices. I will next give my recommendations for a few major types of devices for different situations.

My recommendations for Streaming Device Choices

Cheapest route: <u>Use preexisting used devices</u>. A Laptop or old computer can be kept near the TV and used for entertainment. A HDMI wire connection is preferable, although component, VGA, and S-Video connections are also doable. The only caution is sometimes older computers can't play streaming video smoothly, so a trial is needed prior to setting things up. Go to Hulu.com and try some of the streaming content first.

Next, if you have a SmartTV then your content is already build in. You can activate and use those services.

If you have a Wi-Fi Blue-ray player, then you are already to get service. Activate and use those services.

If you have one of these video game consoles you can use it without additional cost, except the Xbox and other subscription costs.

Best overall device: <u>Roku 3</u> or <u>a Laptop hooked up to your tv</u>. The amount of content is more massive than any other device except a computer.

Fastest current device: <u>Amazon Fire TV</u>. The quick voice search and responsiveness of the device makes this a great device if speed is important to you. Check the device content for its limits to make sure it covers all of your desire content.

Best device for older TVs: <u>Roku 2</u>. Especially if you have a TV that can use composite connections. It has the most content and Wi-Fi remote at a 600 MHz CPU speed.

Best device for iTunes content users: <u>Apple3</u> TV. If you have already bought content for you iPhone, iPad, or iPod, then this is the device that you want.

Best device for modifying: <u>G-Box Q Kodi</u>, <u>G-Box Midnight MX2 XBMC</u> or Android Mini PC device.

Best device to hook up to hard drives: First, <u>WD TV Play</u>, <u>WD TV Live</u> or WD TV Live Plus. Second, VIZIO Co-Star or VIZIO Co-Star LT (although for cord cutters the integration pay TV features will not work on the VIZIO).

Best device USB dongles:

Content - <u>Roku stick</u>, *Speed*- <u>Amazon Fire stick</u>, *Customization*- Android Mini-PC stick, *Price*- <u>Chromecast</u> or <u>Amazon Fire stick</u>

Streaming Devices

Figure out if you already have a streaming device that you can use to connect your television to internet content. Using a preexisting device is the most economical way of trying out streaming, without adding extra cost. You can purchase a device if getting one is absolutely necessary, or a luxury. However, when trying to save money, try to use a device you already own first.

Streaming Devices you may already own

 Advantages: Low cost, you may already own

 Disadvantages: Bulk, not pleasing to the eye

 Cost: free up to HDMI cable $20 (or other device to TV connector)

Computer, tablet, smartphone, laptop

Hook up existing computer, tablet, smartphone, or laptop to Television preferable via HDMI. It will mostly depend on what connectors the television and the computer have built in.

Return to Table of Contents

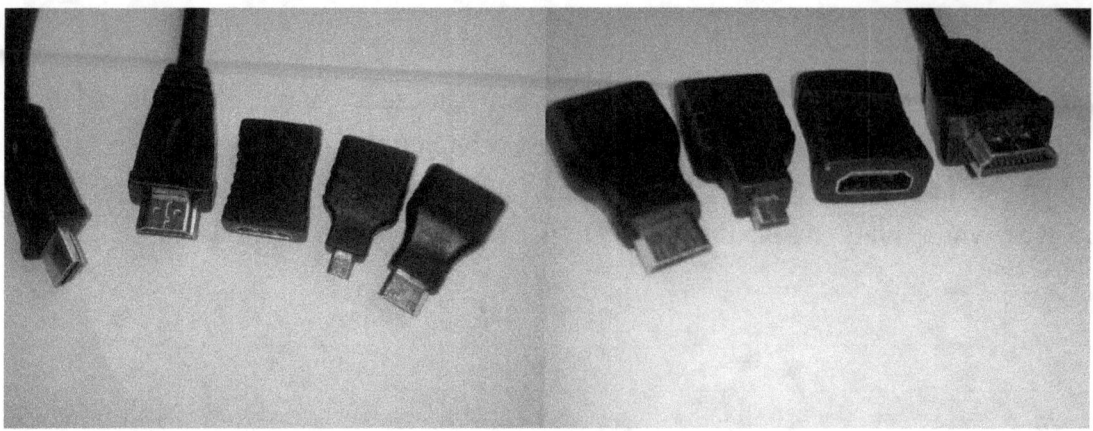

Different HDMI connectors. HDMI regular size, HDMI Mini (Tablets probably use most), HDMI Micro (Smartphones).

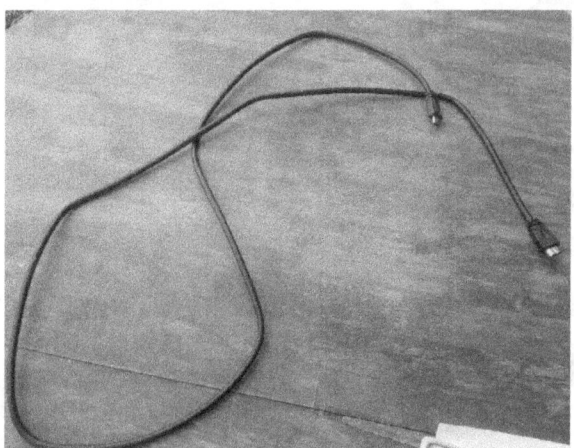

HDMI Wire

The biggest drawback is the inconvenience of having to hook up and unhook the computer if it is not dedicated to the television. Also the computers can be big, bulky, and awkward, in addition to not fitting in with the overall appearance of most entertainment systems.

The tablets and smartphones I have hooked up may get quite warm and use up power quicker if plugged into a charger.

Change TV mode

Normally when you connect your computer, tablet, console, or streaming device to your computer you will need to change the mode or source for.

Return to Table of Contents

	Price	Hulu Plus	Netflix	Amazon Instant	HBO GO	YouTube	Crackle	Popcornflix	VUDU	MGO	iTunes	CinamaNow	PLEX	EPIX	Google Play	MLB TV	NBA TV	NHL	MLS	Internet Browser	Music	Skype	WWE
Xbox	$179 to $287 XBoxes, $60 annual Live Gold membership	✓	✓	✓		✓	✓											✓	✓	✓ Internet Explorer	✓Xbox Music, Muzu TV, Rhapsody, Last FM, iHeartRadio	✓	✓
Wii	$100 to $400	✓	✓	✓		✓	✓														✓Tuneln		
PlayStation	$200 to $600	✓	✓	✓		✓	✓		✓	✓						✓	✓	✓					
Smart TV (various)	$200 +	✓	✓	✓	✓	✓	✓		✓	✓		✓				✓	✓	✓					
Wifi Bluray (various)	$50 to $200	✓	✓	✓		✓			✓	✓		✓				✓					✓Pandora		
Chromecast	$35	✓	✓	✓		✓														✓Chrome Cast	✓Pandora		✓
Amazon Fire TV	$39 to $99	✓	✓	✓		✓	✓		✓	✓				✓					✓		✓Pandora, iHeartRadio, Tuneln		
WD TV Live	$84.99	✓	✓	✓		✓	✓		✓												✓Pandora		
Roku	$49.99 to $94.99	✓	✓	✓	✓	✓	✓	✓	✓			✓	✓	✓	✓	✓	✓	✓	✓		✓Amazon Cloud music, Tune In, iHeart Radio, Slacker, AccuRadio		✓
Apple TV	$92.95 to $114.99	✓	✓			✓					✓				✓	✓	✓	✓			✓iTunes		✓
NETGEAR NeoTV	$43.99 to $74.85	✓	✓	✓		✓			✓												✓Pandora, Rhapsody		
D-Link MovieNite	$40.47	✓	✓			✓			✓												✓Pandora		
RCA Streaming Player	$59.47		✓																		✓Pandora		
VIZIO Co-Star	$58 to $86																				✓Pandora, iHeartRadio		

Check box for exact apps

Video Game Consoles

Advantages: You probably already own it

Disadvantages: Some consoles like Wii have very limited content.

Cost: No more cost if you already own it, just cost of subscriptions. Xbox currently has gold membership annual requirement cost.

Many People already enjoy gaming on video game consoles. Don't buy these units just for their streaming ability. They are priced extremely too high if bought only for their streaming ability. Buy them for their video game content first. Secondary is the ability to play Blue-rays and DVDs. Third is the ability to stream. The three main consoles have the ability to stream through apps.

Xbox

Prices and pictures

360 $163 to $300

One $350 to $600

Content

Current streaming apps include Hulu Plus, Netflix, Amazon Instant, HBO GO, ESPN, Encore Play, FX Now, Fox Now, History, Lifetime, Showtime Anytime, Starz Play, VUDU, Target Ticket, EPIX, Red Bull TV, NBA Game Time, Popcornflix, Crackle, SnagFilms, iHeart Radio, MTV, Dailymotion, and YouTube. To see the apps that you can use to stream see the official 360 or One site. Most apps require additional subscriptions fees in order to access content. http://www.xbox.com/en-US/LIVE/partners

Speed

Xbox 360 has a 3.2 GHz triple core 512 MB RAM, with a 500 GHz GPU. Xbox One has a 1.75 GHz eight core 8 GB RAM, with an 853 MHz GPU. The 360 is said to have problems overheating sometimes. The One is built with cooling in mind with a 10 year lifespan constantly powered on.

Setting it up

1. Xbox with the Xbox live gold membership ($50 currently) and Wi-Fi internet you can stream several streaming services to your television. You may have to pay additional for paid streaming services such as Hulu Plus monthly. Purchase your desired streaming service first from their website. Write down your username/email and password(s)
2. Xbox Live Gold Membership http://www.amazon.com/Xbox-LIVE-Month-Gold-Membership-One/dp/B0029LJIFG
3. Sign in with your Xbox Live Gold Membership enabled gamertag.
4. Go to the App section.
5. Select *Browse Apps*. Select the desired video streaming app(s) to download them. Limit is your memory space.
6. To start the app go to *Video* or *TV & Movies*, then go to **My Video Apps**.
7. Select the one you wish to start.

Wii

Price and picture

Wii $129
Wii U $250

Content

Amazon Instant Video, Hulu Plus, Netflix, YouTube

Speed

Wii has a 729 MHz processor and 512 MB RAM, with a 243 MHz GPU. Wii U has a 1.24 GHz triple core and 2 GB RAM, with 1GB of the 2GB RAM shared between the CPU and the GPU.

Setting up internet connection

1. Power up. Press the "A" button on the remote to access the main menu. Select the "Wii" button with the Wii remote.
2. Select "Wii settings".
3. Select the right arrow ">" to move to page two.
4. Select "internet".
5. Then select "connection settings".
6. Choose wired or wireless instructions based on your situation.

A) Wireless

1. Click on "wireless connection"
2. Make sure there is a clear spot, clear one, and select "Connection 1: None".
3. Next press "Search for an Access Point".
4. Press "OK".
5. Search for your wireless connection.
6. Select your wireless connection.
7. Enter password or key, then select "OK". If it connects you are connected. If not, troubleshooting is needed.
8. Click "Save Settings" and then "Yes"

B) Wired

1. Install Wii LAN adapter connected to the USB port.
2. Connect one end of the Ethernet cable into the LAN Adaptor and the other into the Modem or Router.
3. Click on "wired connection".
4. Select "OK".
5. Wii will attempt to connect to the internet. If successful, press "OK". If not, some troubleshooting is needed to figure out why the system will not connect.

Make sure the system is updated

1. From the main menu of the "Wii Settings"
2. Select the right arrow ">" twice to move to page three.
3. Select "Wii System Update."
4. Select "Yes".
5. Review system update message. Press "I accept" if you agree with terms of the message.

Download the app

1. Select the Wii Shop icon from the Wii Menu.
2. Select "Start Shopping".
3. Select the desired app.
4. Select "yes, keep running," if present with a memory space message.
5. Select "No" if you are not a current subscriber, select "Yes" you already are.

Playstation – PS3, PS4, PS VITA

Prices and pictures

$199 to $270 PS3

$399 to $500 PS4

$200 PSVITA

Content

Current streaming apps include Hulu Plus, Amazon Instant, Netflix, HBOGO, VUDU, EPIX, MLB TV, NBA Game Time, NHL, WWE Network, TuneIn Radio, Crackle, and YouTube.

Also now available is Playstation Vue Pay TV Channels. See section for details, availability, and pricing.

Speed

PS3 is reported to have a 3.2 GHz CPU and 256 MB RAM, with a 550 MHz GPU. PS4 has a 1.6 GHz 8 core shared with APU and 8 GB RAM, with an 800 MHz APU. PSVITA is reported having a 2 GHz 4 core processor and 512 MB RAM, with a 200 MHz 4 core GPU.

Internet connection [Easy mode]

1. Setting up internet connection
2. First make sure you have your access device name and password/key before starting for your wireless connection.
3. Power on your modem and router (if you have one). Wait until all lights required lights are on especially the internet and Wi-Fi.
4. In the XBB main menu go to Settings (icon looks like a tool box).
5. Go to *Network settings* and press the **X** button.
6. Go to *Internet Connection* and select [Enabled].
7. Scroll down to *Internet Connection Settings* and press the **X** button.
8. Select the [Yes] button if it says it will be disconnected.
9. Select [Easy] and press the **X** button

*Follow the wireless or wired procedure.

A) Wireless

Pre start up: Make sure no Ethernet is connected to the PS.

1. At the *Connection Type* screen, select [Wireless] and press the **X** button.
2. At the *WLAN Settings* screen, select [Scan] and press the **X** button.
3. From the list, find your SSID (for modem or router), select it and press the **X** button.
4. Confirm your SSID then press your right button. Don't press **X** button at this moment, which will edit your SSID.
5. Select your Security Type, which is normally WPA or WPA2.
6. Enter your security key or password for the modem or router.
7. Press the right direction arrow.
8. Once all the info is entered, a list of settings will be shown. Press the **X** button to save these settings.
9. Press the X button to test the connection.
10. The test will succeed or fail. A success means that the wireless connection is ready.

B) Wired

Pre start up: Before starting make sure the Ethernet is connected to the PS and the router or modem.

1. After pressing the X button from step 9 of the Easy connection mode, the system will check for your configuration.

2. A list of settings will appear.
3. Press the **X** button to *Save* the settings.
4. Next *Test* the connection by pressing the **X** button.
5. The connection will succeed or fail. A success means you are connected.

Using a Streaming Service

1. If the service requires a subscription, first go subscribe to the service using their website. Take note of any username and passwords.
2. Some free services require at least registration.
3. Connect your PS to the internet.
4. Sign on to your PlayStation Network account. You may have to register if you have not done so previously picking your username and password.
5. Go to your **XMB** (*Xross Media Bar*).
6. Go to your *PlayStation Network* section on the bar.

Installing Streaming App

7. Go to the XMB and find the TV/Video section.
8. Download the desired app.
9. Sign in to the app with your username/email and password.
10. Enjoy

Smart TV

Possible Brands: Samsung, VIZIO, Sony, LG, Panasonic, SHARP, TOSHIBA

Price range

$170 to $3000+

Advantages: All-in-one, compact
Disadvantages: Higher cost, limited apps

Content

Most Smart TVs have: *Amazon Instant Video, *Hulu Plus, *Netflix, **VUDU, Crackle, **M-Go, YouTube, *requires paid subscription. **requires pay for each movie or episode

Setting up

Wi-Fi connection: Just need to hook it up to your network. Several services require subscriptions indicated below.

Blu-ray Player

Wi-Fi enabled Blu-ray players with apps.

Samsung, VIZIO, Sony, LG, Panasonic, SHARP, TOSHIBA

Price range

$50 to $250

Content

Like the Smart TV, most Wi-Fi Blue-ray Players have:

** Amazon Instant Video, *Amazon Prime, *Hulu Plus, *Netflix, **VUDU, Crackle, **M-Go, YouTube,

*requires paid subscription

**requires pay for each movie or episode

Buy a Streaming Device

Chromecast

http://www.google.com/chrome/devices/chromecast/

Price and picture

Chromecast by Google ($29.99) + Chrome Browser (Free)

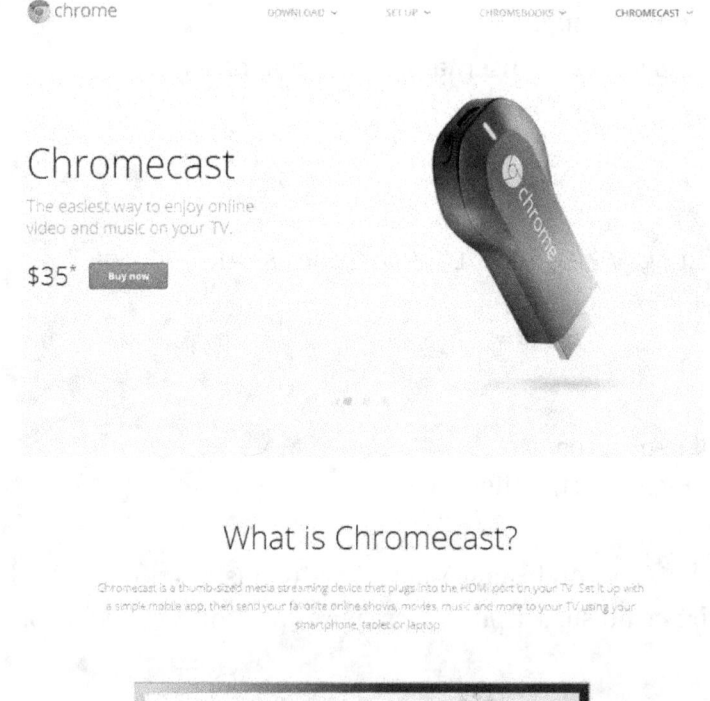

Streams many apps that you cast from your phone, tablet, or Chrome browser on computer. Casting means you send your current internet position to the Chromecast device connected to your TV which it retrieves.

Advantages: Low cost, small, some Chrome casting ability. Pretty improved content.

Disadvantages: It lacks some major content providers such as Amazon, iTune, Sling TV, and has pretty weak games,

Content

*Netflix, YouTube, *Hulu Plus, Crackle, Pandora, iHeartRadio, Rdio, TuneIn, MTV app, Qello Concerts, Google Play Movies and Music, PlutoTV, *HBOGo, *Starz Play, *Showtime Anytime, EPIX, VUDU, SnagFilms, TED, Filmon, Dailymotion, VEVO, *MLBTV, MLS Matchday, MLS Live, NFL Now, Cricket 2015, *WatchESPN, *UFCTV, PAC 12 app. RedBullTV, DramaFever, EROS Now, Comedy Central, NPR One, Watch ABC, Watch Disney, *requires paid subscription

Speed

Chromecast is said to have a 1.2 GHz CPU and 512 MB RAM + 2GB Flash, and Vivante GC1000 GPU.

Setting it up

1. Install the correct version of Google Chrome browser for your operating system.
   https://www.google.com/intl/en/chrome/browser/

2. Open your chrome browser.
3. Search for "chrome extensions" in the browser or go directly to the extension page at
   https://chrome.google.com/webstore/category/extensions .
4. You search for the "Google Cast" in the web store search area.

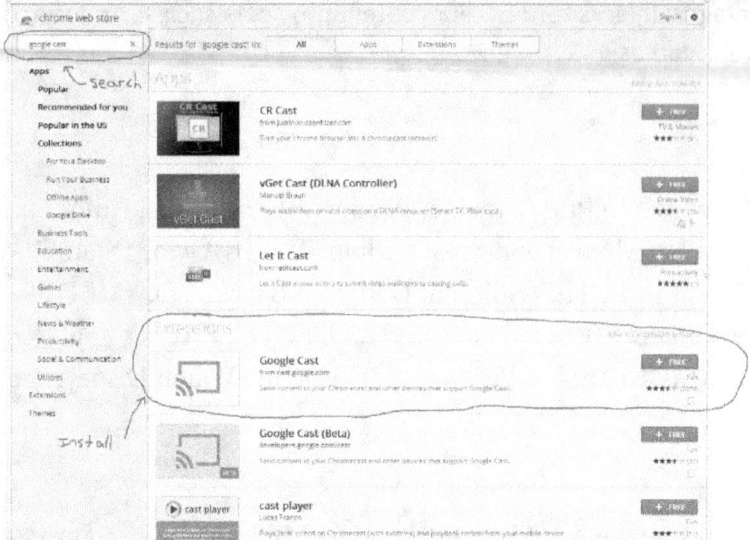

5. Install the free Google Cast extension from cast.google.com.

Hooking up Chromecast

1. Plug the Chromecast into a HDMI port in your television.
2. Plug the other end into either a USB port in your television or into the included power charger leading to a wall socket or power grid.
3. Power on the television.
4. Select the appropriate HDMI as the source or mode of your television.
5. With your chrome browser that you installed, go to the website http://google.com/chromecast/setup with your laptop, phone, or computer when it tells you to.
6. It will also tell you what your Setup Name is: _____
7. You will **download** the Chromecast application. **Install** it.
8. Start the application you downloaded. It will search for your Chromecast device.
9. When it finds your device, or a list of Chromecast devices select yours that you will be using for this TV and **Continue**.
10. Letters and numbers should now appear on both your TV and computer/laptop/phone.
11. Click on **That's My Code** to continue.
12. It will then ask for your network name and password for your wireless network.
13. Then you name your Chromecast device
14. Click **Continue**.

To Cast from the Chrome browser

1. Open the browser. Press the cast button in the browser.
2. Click it again if you want to pause supported videos or music, or disconnect.

Amazon Fire TV & Fire Stick

www.amazon.com/FireTV

Price and picture

$99

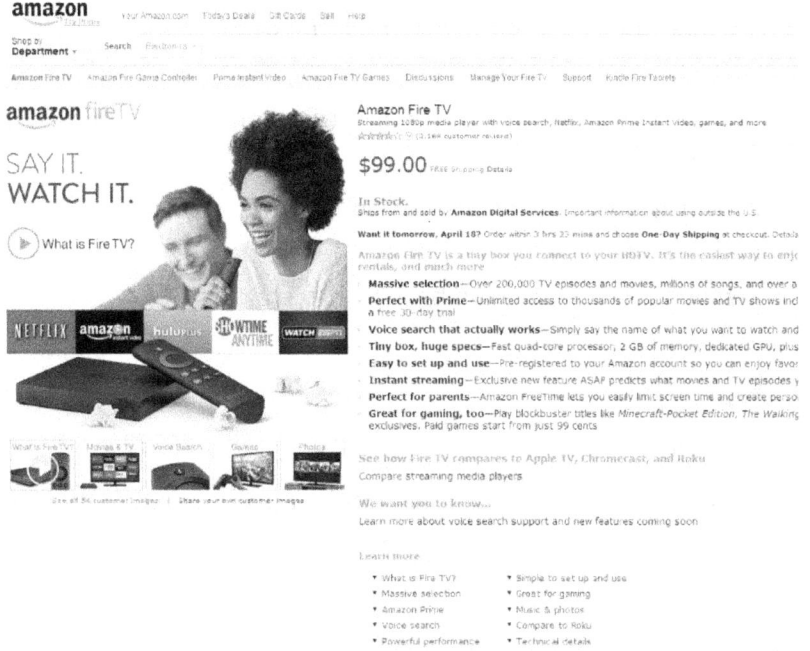

Fire TV Stick - $39

New from amazon is an integrated streaming device for TVs.

> Advantages: High speed, voice search,
> Disadvantages: Higher cost, It lacks a few major content providers such as Google Play, VUDU, iTune, and some sports packages

Contents

Netflix, Amazon Instant Video, Amazon Prime Instant Video, Hulu Plus, Flixster, Showtime Anytime, YouTube, Crackle, RedBullTV, PLEX, Pandora, HBOGo, TuneIn, Spotify, iHeartRadio, Amazon Music, Qello Concert, PlutoTV, TubiTV, IndieFlix, NFL Now, MLB TV, Watch ESPN, NBA Gametime, FOX Sports Go, WWE, Weather Live, Weather4Us, History Channel, A&E, Sling TV, Lifetime, Smithsonian TV, CBS News, FOX News, BBC News, HUFF Post Live, WSJ Live, NEWSY, SKY News, TED, Spotify

Speed

Amazon Fire TV has a 1.7 GHz quad core CPU and 2 GB RAM, with a 400 MHz GPU. It has 8 GB Storage

Amazon Fire TV Stick has a 1.0 Ghz dual core CPU and 2 GB RAM, with a VideoCore GPU. It has 8 GB storage.

Setting up

1. Plug one end of your HDMI into the back of the TV.
2. Plug the other end into your device.
3. Plug the power adaptor into your device. Plug the other in the wall.
4. *if using a Ethernet wire plug it into your modem, then the other into your device
5. A light will come on indicating power is being received.
6. Power on your TV. If your TV shows the Amazon Fire TV logo, then your have everything hooked up ok. If not, check to make sure your input is set to the correct HDMI source.
7. It will search for your remote, so make sure the batteries are in it ready to go.
8. When it finishes it will say press **Play** or **Pause** to start.
9. The system will update the software.
10. It will play an introductory video.

Remote

The remote is very easy to use with voice a voice search button build in. You can also download the Fire TV Remote app and use your smartphone.

G-Box Q & Midnight MX2 XBMC
Price and picture
$88

G-Box Q with Kodi Media Center $109

Advantages: Flexibility and customization, Speed
Disadvantages: High cost, Complexity of installations

Content

Netflix, Hulu Plus, Crackle, YouTube, Pandora, TuneIn, Google Play, CNN, Google, Chrome, Firefox, Twitter, Facebook, Skype, Yahoo, ESPN Scorecenter, PlutoTV, HBOGo, ShowtimeAnytime, WWE Network, Disney Movies, The CW, Watch ABC, FOXNow, StarzPlay, M-GO, EPIX, RabbitTV, entire Google Play App selection.

Speed

The MX2 has a 1.6 GHz Dual core and 1 GB RAM, with 500 MHz quad core GPU. It also has an 8 GB storage Flash.

The Q has a Amlogic Quad Core and 2 GB RAM, with a Octocore GPU. It has 16 GB storage Flash.

Setting up

1. Hook up the HDMI or Composite Video cable(s) to the TV. Connect the other end to the device.
2. Connect the power supply to the device, then plug the other end into the wall or power grid. The device will boot automatically once power is connected.
3. Ethernet connection may be needed HD streaming capability if buffering becomes problematic. Wi-Fi can be connected if you have fast internet.
4. Most customizer users prefer the standard **Launcher** when using the device, over the **3D** or **XBMC** launchers. The standard looks similar to any Android Tablet or Smartphone launcher, so most are already familiar with that UI. It will allow the easiest adding of Google apps from the Play Store.
5. First configure your network connection. To get there click on the 6 dots on top right of the screen (app list button). Click on **Settings**.
    a. If you are going to use Wi-Fi. Slide Wi-Fi slider **ON.** Select your access point, then enter the network password.
    b. For Ethernet connection, enable the Ethernet mode by sliding the Ethernet slider **ON**.
6. You may be prompted to install new updates. If so, do so.
7. You can then run **XBMC** media center and try some of the add-ons. For more information consult XBMC websites and tutorials.
8. You can also run the **Android apps** such as Netflix or download more from the Play store. Consult individual app store information for further details. Also check out the App lists later in the Android device section of this book for a list of specific Android Apps that can be used for your TV.

The Most popular add-ons for the MX2 are Fusion and Xfinity Installers, Mash Up, Ice Films, Navi X, 1 Channel.

Roku 1, 2, 3, LT, Streaming stick
https://www.roku.com/

Roku Streaming Stick $49.99, Roku1 $49.99, Roku2 $64.95, Roku3 $99.99

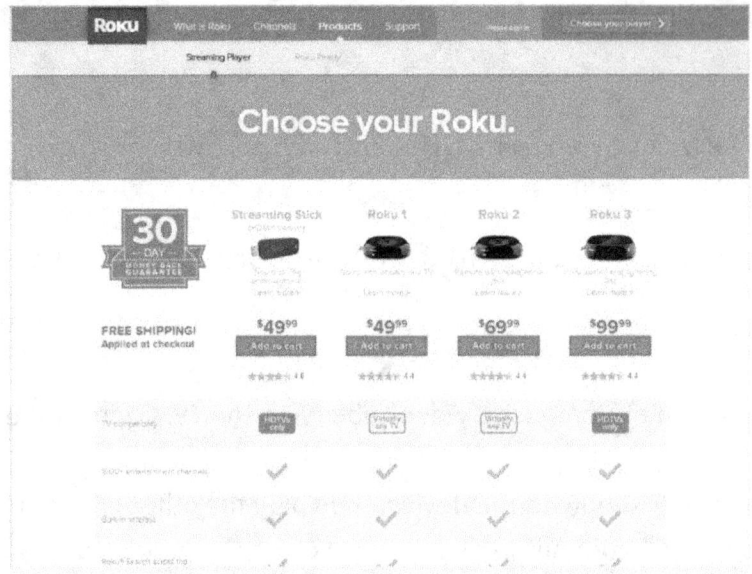

Advantages: Content king, best overall, lower price for slower
Disadvantages: High cost for fastest Roku. RK1 IR remote, Lacks only iTunes content wise.

Content

Roku sports 1,000+ channels in the streaming sense. Most are on-demand. Some are live streaming. Netflix, Hulu Plus, Amazon Instant, GooglePlay, Sling TV, CBS All Access, YouTube, TubiTV, Livestream, VEVO, Crackle, PopcornFlix, SnagFilms, PBS, History, A&E, Lifetime, Smithsonian, Pandora, TuneIn, iHeart Radio, VEVO, Spotify, Rdio, SiriusXM, Amazon Cloud, Slacker, AccuRadio, Qello Concert, PLEX, VUDU, EPIX, MGO, Blockbuster On Demand, Cinama Now, HBOGO, Showtime Anytime, FX Now, Weather Underground, Weather Nation, Weather4US Watch ESPN, MLB.TV, WWE Network, NBA Game Time, NHL Game Center, MLS Live, UFCTV, Crunchyroll, Fox Now, CBS News, NBC News, FOX News, SKY News, ABC News, WSJ Live, Bloomberg, CNBC, NEWSY, RT, HUFF Post live, Newsmax TV, Disney, PBS kids, Popcorn kids, National Geo kids

Speed

Roku 3 has a 900 MHz CPU and is 5 times faster than the Roku 2. **The Streaming Stick, Roku 1, and Roku 2** have 600 MHz CPUs. The Roku 2 varies from the Roku 1 in that the remote is Wi-Fi and the wireless is dual band. So if you get the Roku 1 you may want to connect via Ethernet for best speed since it lacks the dual band Wi-Fi (that is if you have a dual band modem).

Setting up Roku systems

1. Connect HDMI to TV (or composite (SD) for older TVs for Roku 2 if needed).
2. Connect the other end to you Roku Player.
3. Put batteries in the remote (except the Ready-to-go Streaming Stick).
4. Connect power adapter to the Roku, then the other end to the wall.
5. Turn on your TV. If it is in there is no Roku logo, you will need to change the input source to the proper HDMI input mode.
6. Enter your network method. Wi-Fi or Ethernet wire.
7. Roku will tell you to get on to a computer and go to roku.com/link
8. Go there and enter the code shown on the TV on the computer.

Remotes

Roku3 has a Wi-Fi & motion sensor remote. Roku 2 has Wi-Fi. The Streaming Stick (HDMI version) has a RF one. Roku1 has an IR remote (line of sight required).

WD TV Live

Price and picture]

WD TV Live $139.97

WD TV Live Plus $149.98

WD TV Play $91.97

WD TV Media Player $84.93 (lacks Netflix)

http://www.wdc.com/en/products/products.aspx?id=1270

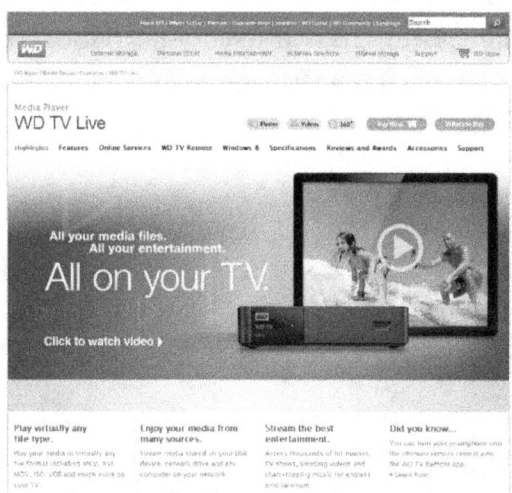

Advantages: Integration with hard drive content, covers major content
Disadvantages: older, High cost, lacks Amazon, lacks sports content

Contents

Netflix, Hulu Plus, YouTube, Pandora, TuneIn, Spotify, Daillymotion, Vimeo, VUDU, Cinema Now, Facebook, AccuWeather, MLB.TV Premium, SnagFilms, RedBull TV

Speed

700 MHz CPU and 512 RAM, 256 Flash Memory.

Year

2011

Setting up

1. Connect the included composite cable up to the TV. Or connect an HDMI cable (not include, but recommended over composite for newer TVs) up to the TV.
2. Connect the other end to the device in the proper port.
3. Plug in the other end to an outlet or surge protector. Connect the other end to the device.
4. Connect a USB storage device to the device if you will be using an external drive.
5. Press the power button on the remote. The indicator light will come on.
6. Turn on your TV. If you are not in the correct source, change the TV mode to the proper input source.
7. Select your language.
A. Connect your device to your network via Ethernet for wire connection. The wizard will begin the automatic connection. When finished press **OK**.
B. For wireless press up and down to find your access point. Press **OK** to select your service. Use the arrow direction buttons to type in your password. Then hit the submit screen button when finished. Press on the **OK** remote button to continue.

Apple TV

https://www.apple.com/appletv/

Price and picture

Model MD199LL/A $89.99

3rd Gen $96.99

Advantages: Only iTunes content specialization, has major content
Disadvantages: High cost, Lacks Amazon, Google Play, VUDU, MGO, Popcornflix, Sling TV

Content

Netflix, Hulu Plus, Crackle, YouTube, VEVO, Vimeo, Dailymotion, Qello Concert, HBO GO, HBO Now, Showtime Anytime, FX Now, A&E, Lifetime, FYI, History, PBS, TED, Smithsonian, Watch ABC, Fox Now, Watch Disney, Smithsonian, Red Bull TV, The Weather Channel App, Watch ESPN, MLB.TV, NFL Now, NBA Game Time, NHL Game Center, MLS Live, UFCTV, WWE Network, ABC News, CNN Go, CBS News, CNBC, Bloomberg, Sky News, WSJ Live, Crunchyroll

Speed and resolution

Apple TV 2nd gen 750 MHz and 256 RAM. 720P. **Apple TV 3rd gen** 1 GHz and 512 RAM, 400 MHz GPU. 1080p.

Setting up

1. Connect a HDMI (not included) cable to your TV. Connect the other end to your device.
2. Connect the power supply to the wall. Connect the other end to your device.

3. Turn your TV on. If you don't see the Apple information, make sure your TV is set to the proper input source mode.
4. You should see the Apple logo and language selection screen. Use the direction pad to select the proper language, press the center button to select that language.
5. It will then go to your Wi-Fi setup. Find your access, and enter the password.
6. Choose whether to help Apple with information or not.
7. To use Apple TV with your other computer iTune content go to **Computer** to set up *Home Sharing* option

NETGEAR Neo TV, Neo Max

http://www.netgear.com/landing/stream/tv/#neotv

and pictures

NETGEAR Neo TV (NTV300) (2012) $35.66

NETGEAR Neo TV MAX (2012) $39.99

NETGEAR Neo TV Prime (2012) $65.99

NETGEAR Neo TV Live (2011) $79.99

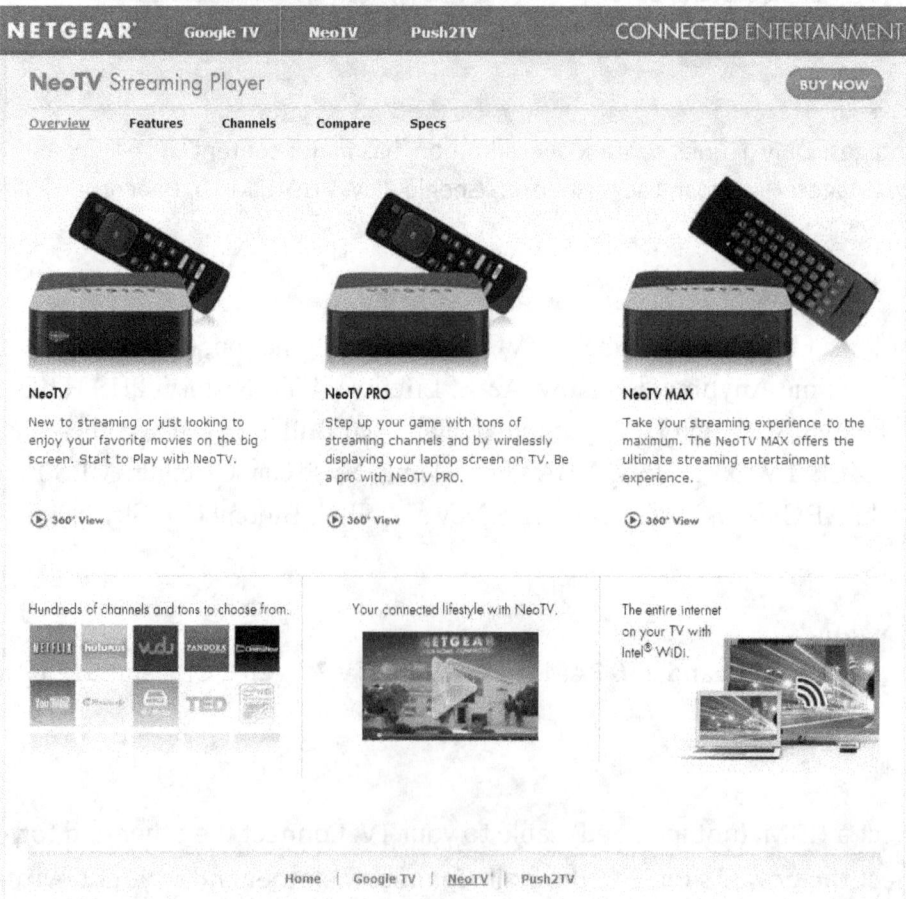

Advantages: lower cost, covers basic streaming
Disadvantages: Lacks major Amazon, iTunes,

Contents

Netflix, Hulu Plus, YouTube, Pandora, Rhapsody, VUDU, Cinema Now, TuneIn, Spotify, TED, TV Guide

Speed

1080p NeoTV Max Mediatek MT8653 CPU and 512 RAM, 256 Mb Flash

Remotes

The Prime and Max remotes have QWERTY keyboards. The Prime model has a Bluetooth remote. The rest are IR remotes.

D-Link MovieNite & Movie Nite Plus

http://us.dlink.com/product-category/home-solutions/entertain/media-players/

Price and picture

$29 MovieNite DSM-310

$39 MovieNite Plus DSM-312

Advantages: Low cost, basic streaming services
Disadvantages: Limited content, Lacks most major content providers

Content

Netflix, YouTube, Vudu, Pandora, Picasa

Setting it up

I won't go into details. These are less popular systems. It is basically the same network info and account info as the others. So see the other device setup if help is needed.

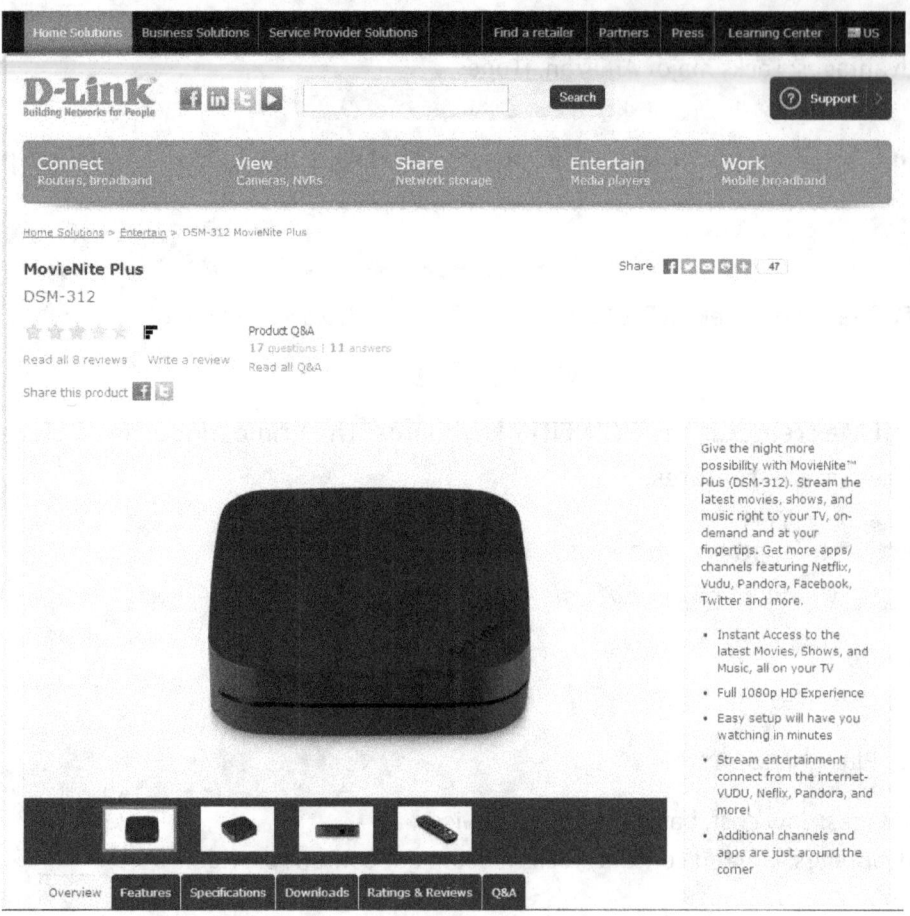

RCA Wi-Fi Streaming Media Player

Price and Picture

$59.60 DSB876WU-WH

$99.99 DSB772E

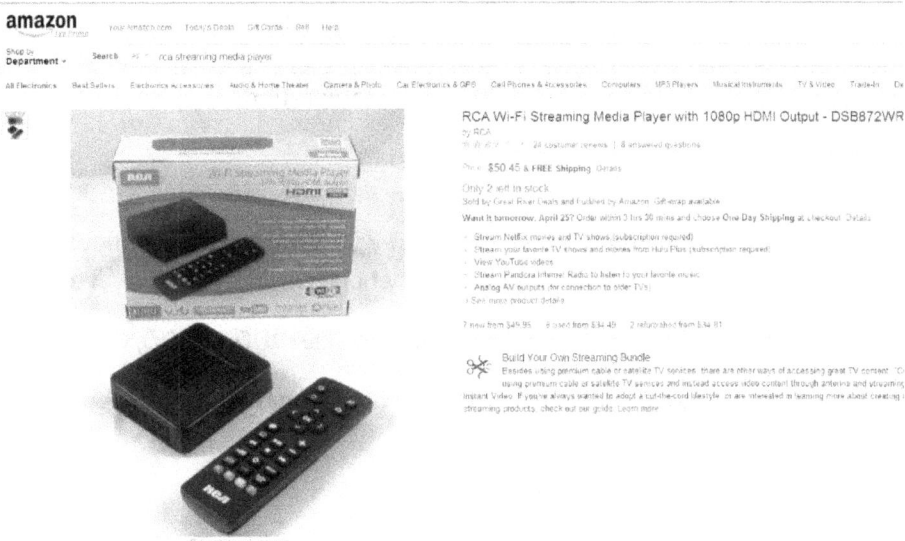

Advantages: basic streaming

Disadvantages: this is a poor streaming at too high price, get something else

Content

Netflix, Hulu Plus, YouTube, VUDU, Pandora, Picasso

VIZIO Co-Star and Co-Star LT

Price and pictures

$89.99Co-Star LT

$99 Co-Star

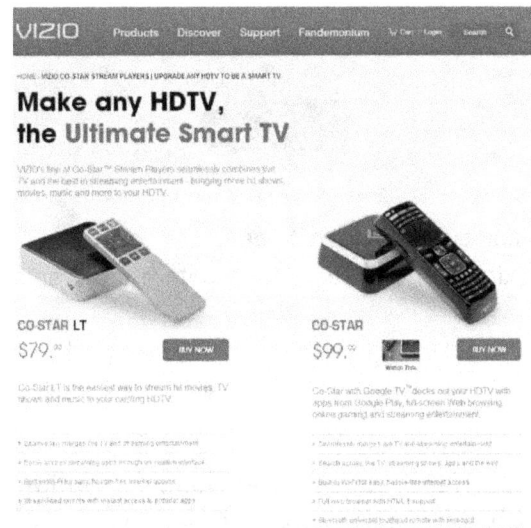

Advantages: integration of Pay TV and hard drive, mid level content availability, mid level speed, full keyboard on Co-star (not LT)
Disadvantages: cord cutters don't need the pay TV integration

Content

This is made to set up with cable or satellite pay TV. So, it will not integrate as it does with cable/satellite sadly with the free OTA TV antenna service. It will still function as a streamer however.

Netflix, Hulu Plus, Amazon Instant, Google Play, YouTube, Crackle, VUDU, MGO, Pandora, iHeart Radio

On live app turns the system into a video game console with Batman Arkham, Lego Hobbit, Lego Movie, Mortal Combat, Metro, and Darksiders II

Co-Star has Chrome Web Browser, LT doesn't.

Speed

1.2 GHz dual-core processor, with a 750 MHz GPU. 1 GB RAM. 4 GB Flash.

Remote

LT is IR. Co-Star has Bluetooth remote with touchpad and built in keyboard on backside.

Android Mini PCs

Popular the last few years is the Android Mini PCs. These devices essentially turn your TVs into giant tablet computers without the touch screens.

Price and picture:

Advantages: Flexibility and customization, Speed, Low cost
Disadvantages: Complexity of installations, no remote (though smartphone app remote)

What to look for in specifications

Android OS

Newest 4.4 **KitKat** – User interface appearance changes, optimized for low RAM devices, wireless printing, select default text messaging and home launcher app. 5.3% Android devices. For more features see: http://developer.android.com/about/versions/kitkat.html

Most Popular 4.1 to 4.3 **Jelly Bean**

4.3 - AV Bluetooth remote support, OpenGL ES 3.0 support. 8.9% Android devices. For more features see: http://developer.android.com/about/versions/android-4.3.html

4.2 - Lock screen improvement, "Daydream" screen saver when docked, Miracast support, blind user support, new clock/timer, group SMS, Bluetooth gamepads and joystick support (4.2.1), 18.1% Android devices. For more features see: http://developer.android.com/about/versions/android-4.2.html

4.1– Smoother UI, bi-directional text (for multi languages), user installed keyboard maps, turn off notifications per app, rearrange + resize widgets, Bluetooth data transfer Android Beam, offline voice dictation, better camera, better voice search, USB audio supported. 34.4% Android devices. For more features see: http://developer.android.com/about/versions/android-4.1.html

4.0 **Ice Cream** - Visual voicemail, screen capture (power button + volume-down button), lock screen app access, better copy/paste, continuous voice dictation, face unlock, shutdown background data apps, improved camera, new gallery layout, new photo editor, Wi-Fi Direct, Android Beam near field communication. 14.3% Android devices. For more features see: http://developer.android.com/about/versions/android-4.0-highlights.html

3.0 **Honeycomb** – New UI, simplified multitasking w/ snapshot, new keyboard, simplified copy/paste, multi browser tabs, multi core processor support, encrypted data, USB accessory connection, external keyboard and mouse support, joystick and gamepad support. For more features and details see: http://developer.android.com/about/versions/android-3.0-highlights.html

2.3 **Gingerbread** – VoIP support, copy/past improvement, NFC, Plus all the updates from the prior versions which will not be shown here. 17.8% Android devices. For more features see: http://developer.android.com/about/versions/android-2.3-highlights.html

Android statistics are from developer.android.com (Android OS, 2014)

CPUs & GPUs
The CPU performs most app tasks and processing. The GPU performs the graphic processing. More cores allow more multi-tasking of processing. Bolded is the top current processers.

ARM Cortex A8 -OMAP 3 600 MHz mostly to 1.2 GHz, GPU PowerVR SGX530 – 277 to 384 MHz
Snapdragon S1 – 528 MHz to 1 GHz, GPU Adreno 200 – 200 MHz
Snapdragon S2 – 800 MHz to 1.5 GHz Dual Core, GPU Adreno 205 – 333 MHz
Snapdragon S3 – 1.7 GHz Dual Core, GPU Adreno 220 – 500 MHz

Snapdragon S4 – 1 to 1.7 GHz Dual and Quad Core, Adreno 203, 225, 305, & 320 - 500 to 533 MHz

Snapdragon 200 – 1.2 to 1.4 GHz Dual and Quad Core, Adreno 203 & 302

Snapdragon 400 – 1.2 to 1.4 GHz Dual and Quad Core, Adreno 305 – 450 to 533 MHz

Snapdragon 600 – 1.5 to 1.7 GHz Quad Core, GPU Adreno 320 – 400 MHz

Snapdragon 800 – 2.26 GHz Quad Core, GPU Adreno 330 - 450 MHz

Exynos 3 – 1 to 1.2 GHz, GPU IT PowerVR SGX540 - 200 MHz

Exynos 4 – 1.2 to 1.6 GHz, Dual and Quad Core, GPU ARM Mali-400 MP4 - 266 to 522 MHz

Exynos 5 – 1.7 to 2.1 GHz, Dual and Quad Core, GPU ARM Mali-T628 & 624 MP6 or IT PowerVR SGX544MP3 - 480 to 695 MHz

Exynos 5420 - 1.9 GHz octa-core, a six-core ARM Mali-T628 GPU 695 MHz

Tegra 2 – 1 to 1.2 GHz, Dual Core, GPU 300 to 400 MHz

Tegra 3 – 1.2 to 1.6 GHz, Quad Core, GPU 416 to 520 MHz

Tegra 4 – 1.9 GHz, Quad Core, GPU 672 MHz

RAM

RAM is the working memory of the computer device.

256, 512, 768, 1024 (1GB), 2048(2GB)

Internal Storage
16GB, 32GB

Micro SD Card Slot
Varies

Connectivity
Wi-Fi 802.11 a/b/g/n/ac, Bluetooth 2.0 + EDR, Bluetooth v2.1 with A2DP, Bluetooth 3.0, Bluetooth v4.0 HS, A-GPS,

Input
Micro-USB; USB On-the-go, 3.5 mm

Output
Screen
HDMI
Micro HDMI
Mini HDMI

Streaming Apps for Mobile and Mini Devices

Android Apps

Network

NBC ABC FOX (CBS, CW) – see Hulu Plus app Hulu Plus. Android Varies

https://play.google.com/store/apps/details?id=com.hulu.plus

CBS http://www.cbs.com/mobile/   Android 4.0 and above. Episodes

CW https://play.google.com/store/apps/details?id=com.cw.fullepisodes.android Android 2.2 and above. Episodes

ION https://play.google.com/store/apps/details?id=com.iontelevision.mobile . Clips. Android 4.0 and above

Public TV

PBS MHz Network https://play.google.com/store/apps/details?id=com.mhz.MhzNetworks Android 1.6 and above. Live, Episodes

PBS NHK World (Japan) https://play.google.com/store/apps/details?id=jp.nhkworldtv.android Android 2.2 and above. Live

PBS France 24 https://play.google.com/store/apps/details?id=com.france24.androidapp Live and on demand. Android varies

Music TV

Zuus https://play.google.com/store/apps/details?id=com.zuus.app Android 2.2 and above. 50 music genre stations.

Weather

Accuweather https://play.google.com/store/apps/details?id=com.accuweather.android Android 2.2 and above

Weather Nation https://play.google.com/store/apps/details?id=com.xav.wn Android 2.3 and above

Streaming TV

Hulu Plus https://play.google.com/store/apps/details?id=com.hulu.plus

Streaming movies Android

Netflix (subscription required)
https://play.google.com/store/apps/details?id=com.netflix.mediaclient Android Varies

Crackle (free movies) https://play.google.com/store/apps/details?id=com.gotv.crackle.handset Android 2.3.3 and above

VUDU (Pay per)
https://play.google.com/store/apps/details?id=air.com.vudu.air.DownloaderTablet

PRICE COMPARISON CHART 2015

System	System cost (minus other costs)		
Existing PC, Tablet, notebook	Free		
Existing Wii or Playstation	Free		
Existing Xbox	Gold membership $50 annually		
Existing SmartTV	Free		
Existing Wi-Fi Blue-ray Player	Free		
D-Link MovieNite	$29.00		
Chromecast	$31.59		
NETGEAR NeoTV	$35.66		
Amazon Fire Stick	$39.00		
D-Link MovieNite Plus	$39.00		
NETGEAR NeoTV Max	$39.99		
Roku Streaming Stick	$49.00		
Roku1	$49.99		
Older Blue-ray Streaming Player	$50.00		
Older Android Mini PC	$50.00		
Roku2	$64.95		
NETGEAR NeoTV Prime	$65.99		
RCA Streaming Player	$67.95		
NETGEAR NeoTV Live	$79.99		
Newer Android Mini PC	$80.00		
WD TV Media Player	$84.93		
G-Box MX 2	$88.00		
Apple TV	$89.99		
Vizio Co-Star LT	$89.99		
WD TV Play	$91.97		
Roku3	$93.69		
Apple3 TV	$96.99		
Vizio Co-Star	$99.00		
Amazon Fire TV	$99.00		
RCA Streaming Player DSB772E	$99.99		
Mid Grade Blue-ray Streaming Player	$100.00		
Wii	$100.00		
G-Box Q	$109.00		
WD TV Live	$139.99		
WD TV Live Plus	$149.97		
Xbox 360	$163.00		
Playstation3	$200.00		
Low price small SmartTV (various)	$200.00		
New WiFi Blueray (various)	$200.00		
PS Vita	$200.00		
Wii U	$250.00		
Xbox One	$350.00		
PS 4	$399.00		

	Year	Price	Installation	Content	Remote Control	Mobile Remote Apps	Speed	Memory	Blue-ray and DVD
Xbox	360 2005, One 2013	$179 to $500 XBoxes. $50 annual Live Gold fee a minus membership		Great	$20 separate unless comes with bundle	NA	360 3.2 Ghz Triple, One 1.7 8core	4GB to 160GB (XBOne 500GB)	Yes
Wii	Wii 2006, Wii U 2012	$100 to $400	EZ	Limited	controller	NA	Wii 729Mhz, Wii U 1.24 Ghz 3core	512MB + SD card (8 to 32GB Flash)	No
Playstation	PS3 2006, PS4 2013	$200 to $600	Ez	Great	$10 to $40	NA	PS3 3.2 Ghz, PS4 1.6 Ghz 8core, PSVita 2Ghz 4core	20 to 160GB (PS4 500GB), PS3 256 Mb or PS4 + Vita 512 Mb RAM	Yes
Smart TV (various)	Samsung 2012 VIZIO 2012 - 2013 LG 2013 SONY 2013	$200 +	EZ	Good	in TV remote	Sony, LG, VIZIO, and Samsung	2yrs old	Normally under 1GB	If built in
WiFi Blueray (various)	2013	$50 to $200	EZ	Good	Yes	Sony and Samsung	1yr old	Normally under 1GB	Yes
Chromecast	2013	$35	Med	Ok	laptop, tablet, or smartphone	iOS and Android Apps, Chrome Browser	1.2 Ghz	512 Mb	No
Amazon Fire TV	2014	$99	EZ	Good	Yes, with voice search	Kindle Fire	1.7 Ghz quad	2 Gb RAM, 8 Gb	No
G-Box Midnight MX2 XBMC	2013	$98	Complex	Depends	Yes	iOS and Android Apps	1.6 GHz Dual core	1 GB RAM, 8 Gb Flash	No
WD TV Live	2011	$84.99	Complex	Ok	Yes	iOS and Android Apps	700 MHz CPU	512 RAM	No
Roku	Roku 1, 2, 3, LT 2013; Roku Streaming Stick 2014	$49.99 to $94.99	EZ	Great	Yes	iOS, Android, Kindle Fire, Windows, Windowsphone	RK3 900 Mhz, RK2 + RK1 600 Mhz	256Mb	No
Apple TV	Apple TV 2012; Apple TV3 2014	$92.95 to $114.99	Med	Good	Yes	iTunes and iOS	2nd Gen 750 Mhz, 3rd Gen 1 Ghz	2nd Gen 256 RAM, 3rd Gen 512 RAM	No
NETGEAR NeoTV	Neo TV 2011; Neo TV Max 2012	$43.99 to $74.85	EZ	Ok	Yes	iOS and Android	Slow	Neo Max 512 Mb RAM, 256 Mb Flash	No
D-Link MovieNite Plus	2012	$40.47	EZ	Limited	Yes	iOS and Android	Slow	?	No
RCA Streaming Player	2012	$59.47	Poor	Limited	Yes	no	Crawling	?	No
VIZIO Co-Star	Co-Star 2012 Co-Star LT 2013	$58 to $86	Med	Good	Yes	no	1.2 Ghz Dual core	1 Gb RAM, 4 Gb Flash	No
Android Mini PC	RK 3188	$50 to $80	Complex	Depends	No	Android	1.6 Quad core	2 Gb RAM, 8 Gb Flash	No

Streaming Sources

1. Antenna + Hulu free or Hulu Plus subscription + news network website clips, articles, and episodes = rough equivalent to basic cable. Sling TV is also a great option. Playstation Vue is pricey, but an option.

2. Get a Library Card and check out what movies they have and what online streaming services they might have. My library has free FilmsOnDemand and Indieflix, free music, and 50 free magazines from Zinio.

3. Add free Crackle, Popcornflix, Livestream, PlutoTV, TubiTV, SnagFilms, Filmon, and YouTube content.

4. Add to the above an EPIX movie subscription (all have a free trial month) of either Amazon Prime ($99/yr) or Netflix ($7.99) depending on your bonus focus. Amazon Prime includes free 2 day shipping with amazon purchases, Kindle Lending Library, A&E and History channel series like Duck Dynasty and Ancient Aliens. Netflix includes older TV series, foreign, and older movies in addition to the EPIX current first run movies. **Redbox kiosks offers release from their Kiosk $1.50.** You might even want to rotate through the different services though Amazon Prime a year once the trial is completed.

5. If you enjoy CBS shows, CBS All Access is an option.

6. Premium Series not included in the options above you can buy and stream from Amazon, iTunes, VUDU, YouTube, Google Play, CinamaNow, or M-Go normally for SD 1.99 to $2.99/episode or HD 3.99 to 4.99/episode. Only do this for content not included in the free or subscription services above.

7. Add sports entertainment package if you desire. NFL, MLB, NBA, NHL, UFC, WWE. Most include streaming anywhere.

8. Rent Movies subscription Netflix Disk, Redbox Kiosk, or on a per use basis Amazon Instant, iTunes, Google Play, VUDU, CinamaNow, Blockbuster On Demand

9. Prices are determined by what content you enjoy.

Hulu streaming plus CBS through their TV.com streaming services really stands alone together for Network Broadcast TV shows recently played. They also play a few entire series, many times requiring Hulu Plus $7.99 per month. In addition, Hulu also has a lot of basic cable content.

My Recommendations for streaming services

The main consideration for which services to subscribe to Is HuluPlus and which other streaming service subscription should you consider.

Hulu Plus, although it has some movies, really is about TV Shows. There are a few movies.

Largest amount of content: If you are looking for the largest amount of older films and TV shows then Netflix is the best.

If you are looking for new releases: Then Redbox kiosk with their low price kiosk rentals is the best deal. Netflix has their new releases with their DVD plans, though waiting for them can be frustrating sometimes some new releases taking forever to be shipped, and you need to subscribe to their disk mail services. For Redbox, you must live near a kiosk, so check on their website for available content. Even then, Netflix has long waiting lists many times for new release disks.

If you read ebooks and order from Amazon.com: Amazon Prime may be the streaming service for you. In addition, you gain free book checkout service from their Lending library, and free shipping on qualified orders.

Free route: Hulu free (on computer), with library card content, with Crackle, Snag films, PopcornFlix, PlutoTV, TubiTV, YouTube movies, free trial offers, & Filmon movies. Also you can use the list at the end of the book to find free streaming and episodes from several networks.

Streaming Service Providers

Hulu (NBC, ABC, Fox, the CW, ION, WB, BBC, MTV, BET, VH1, Nat Geo, Syfy, Spike, TBS, TNT, USA, Oxygen, Lifetime, LMN, E!, AMC, WWE, Food Network, Disney, Comedy Central, A&E, Bio, DIY, HGTV, History, H2, TVLand, Travel Chan, IFC, Jim Henson, Cartoon Network, Adult Swim, Nickelodeon, PBS Kids, Anchor Bay, Image Ent, Lions Gate, MGM, Anime Net, Aniplex, Funimation, Bandai, Manga, TokyoPop, TOEI, VIZ, Exercise TV, GAIAM TV, Fora TV
http://www.hulu.com/

Hulu has two services 1) free and 2) Plus. Most devices only can receive Hulu Plus. Hulu Plus currently is a free week followed by $7.99

Hulu is a joint venture of NBC Universal Television Group (Comcast), Fox Broadcasting Company (21st Century Fox) and Disney–ABC Television Group (The Walt Disney Company), with funding by Providence Equity Partners, the owner of Newport Television

Movie and TV Library
FOX: FOX, FX, Fox News Channel, Fox Business Network, National Geographic Channel
NBC Universal: NBC, CNBC, MSNBC, Syfy, Oxygen
Walt Disney Company: ABC, ABC Family
A&E Television Networks: The Biography Channel, History Channel, A&E
Turner Broadcasting System/Warner Brothers: TBS, TNT, Cartoon Network, truTV, CNN, Adult Swim, The CW
AMC Networks
ION Television but not the Qubo shows
Food Network: Cooking Channel
WWE shows

Anchor Bay
Additionally, Hulu offers CBS Broadcasting content in Japan and now links the US Original Content

Hulu Original Series
A Day in the Life (2011–2012)
Battleground (2012)
Spoilers (2012)
Up to Speed (2012)
The Awesomes (2013)
Quick Draw (2013)
The Wrong Mans (2013)
Behind the Mask (2013-)
The Hotwives of Orlando (2014)

Popular TV Shows
Modern Family
Law and Order: SVU
South Park
Nashville
New Girl
The Daily Show with John Stewart
Family Guy
The Tonight Show with Jimmy Fallon
Revolution
Naruto
Chicago PD
The Voice
Grey's Anatomy
SpongeBob Squarepants
Glee
Suburgatory
American Dad
Community
The Middle
Bones
The Originals
The Goodwife

Parks and Recreation
Chicago Fire
Trophy Wife
Once Upon A Time
The Goldbergs
Scandal
The Following
Dragonball Z
Bachelor
Arrow
Saterday Night Live
Raising Hope
One Piece
Pokemon
Castle
Sabrina
Late Night with Seth Meyers
The Vampire Diaries
Doctor Who
Jimmy Kimmel Live
Cosmos

Popular Movies
The Usual Suspects
Good Dick
Bad Kids Go to Hell
Steel Magnolias
Lonely Hearts
Husbands and Wives
Babe
Veggie Tales
Resident Evil: Apocalypse
The Girl with the Dragon Tattoo
Dark House
Who's Harry Crumb
Akira
Pokeman
Brutal

Atlantic Rim

Intolerable Cruelty

Lost in Translation

CBS All Access

www.cbs.com/all-access/

One Week free.

Available: Roku, Android, iOS, Kindle, Windows Phone, Blackberry

This is a new service offered by CBS for the price of $6 a month. The service includes next day aired shows, less ads [though not ad free on new shows], and ad free classics archive.

CBS Content

CSI: Crime Scene Investigation
NCIS
Criminal Minds
The Mentalist
NCIS: Los Angeles
The Good Wife
Hawaii Five-O
Blue Bloods
Person of Interest
Elementary
Under the Dome
Extant
Madam Secretary
Scorpion
NCIS: New Orleans
Stalker
Two and a Half Men
The Big Bang Theory
Mike and Molly
2 Broke Girls
Mom
The Millers
The McCarthys
Big Brother
Survivor
The Amazin Race
Undercover Boss
Late Show with David Letterman
The Late Late Show with Craig Ferguson
The Young and the Restless
The Bold and the Beautiful
The Talk

Let's Make a Deal
The Price is Right
60 Minutes
48 forty eight hours

Classic CBS

The Andy Griffith Show
Beverly Hills 90210
Brady Bunch
Cheers
CSI: Miami
Everybody Loves Raymond
Family Ties
Frasier
Ghost Whisperer
I Love Lucy
Macgyver
Medium
Melrose Place
Mission Impossible
Perry Mason
Sabrina the teenage Witch
Star Trek [All series]
Taxi
Twilight Zone [original]
Twin Peaks
Wings

Netflix (EPIX) (Back catalog Time Warner, Universal, MGM, Paramount, Lions Gate, Sony, 20th Cent Fox, Disney)
https://www.netflix.com/

Netflix's "Watch Instantly" service holds first-run rights to films from Paramount Pictures, MGM, Lions Gate Entertainment (through an output deal with Epix), along with back-catalog titles to films from Time Warner, Universal Pictures, Sony Pictures, Paramount Pictures, MGM, Lions Gate Entertainment, 20th Century Fox, Disney, and other distributors. Netflix also provides current and back-catalog TV programs distributed by NBC Universal, 20th Century Fox, Sony Pictures, Disney-ABC Domestic Television, with select shows from Warner Bros. as well. Netflix has "pay TV window" deals with Relativity Media, FilmDistrict, and Open Road Films.

Original Content
House of Cards,

Hemlock Grove, and
Orange Is the New Black
Arrested Development

Netflix received a five-season deal for four Marvel Super Heroes: Daredevil, Jessica Jones, Iron Fist, and Luke Cage. The deal involves the broadcast of four 13-episode seasons that culminate in a mini-series called The Defenders. Broadcasting will commence in 2015.

Netflix Instant Video
Major Studio New Movies (Based on current contracts)

EPIX
Paramount
For a list of recent movies by Paramount
http://en.wikipedia.org/wiki/List_of_Paramount_Pictures_films#2010s

MGM
For a list of recent movies by MGM
http://en.wikipedia.org/wiki/List_of_Metro-Goldwyn-Mayer_films#2010s

Lionsgate
For a list of Lionsgate movies
http://en.wikipedia.org/wiki/List_of_theatrically_released_Lionsgate_films#2010s

Relativity Media
For a list of Releativity movies
http://en.wikipedia.org/wiki/Relativity_Media

FilmDistrict
For a list of FilmDistrict movies
http://en.wikipedia.org/wiki/FilmDistrict

Open Road Films
For a list of Open Road Films movies
http://en.wikipedia.org/wiki/Open_Road_Films

New and Old TV shows (based on major studio contracts)
NBC http://en.wikipedia.org/wiki/Universal_Television

20th Century Fox http://en.wikipedia.org/wiki/20th_Century_Fox_Television
Sony Pictures http://en.wikipedia.org/wiki/Sony_Pictures_Television
Disney-ABC http://en.wikipedia.org/wiki/Disney%E2%80%93ABC_Domestic_Television
Warner Brothers (select shows) http://en.wikipedia.org/wiki/Warner_Bros._Television

Popular TV shows

New Girl FOX– past seasons when season is over

The Walking Dead AMC – past seasons when season is over

Mad Men AMC – past seasons when season is over

How I met your Mother CBS– past seasons when season is over

Family Guy FOX– past seasons when season is over

Archer FX – past seasons when season is over

Clone Wars TOON – current series

Good Luck Charlie DISN – past seasons when season is over

Completed Series

Breaking Bad

Dexter

Amazon Instant Video (EPIX)

Amazon Instant Video is an Internet video on demand service offered by Amazon in the United States, United Kingdom, Germany and Japan. It offers television shows and films for rental and purchase. Except in Japan, a number of titles are available free through to customers with an Amazon Prime subscription.

Amazon instant videos rented, bought, and free through Amazon Prime can download to computers, a few media boxes like TiVo and Roku, a few smart TVs, console gaming units such as PS3, Xbox, and Wii, Apple IPhone and IPad, and Kindle Fire. Check with device specification features to see if Amazon instant videos will work with them.

Amazon Instant Video Membership

Amazon Instant Videos can be purchased or rented on an individual basis. Also Amazon has a Prime membership that includes some free movies, TV shows, Kindle Owner's lending library, and includes some free two day shipping on some qualified items bought at Amazon.com. Currently, Amazon Prime offers a free month trial, and $99 a year.

Movie and Video Content

EPIX (pronounced "epics") is an American hybrid premium cable and satellite television network, and subscription video on demand service that is operated by Studio 3 Partners LLC, a joint venture of Viacom (specifically its subdivision Paramount Pictures), Metro-Goldwyn-Mayer and Lions Gate Entertainment.

HBO and Showtime runs older series now on Amazon available through prime.

Amazon Prime Popular Movies

Star Trek: Into Darkness

The Hunger Games

Marvel's The Avengers

Frozen Ground

The Spectacular Now

Pain & Gain

Jack Reacher

Robin Hood Men in Tights

The Guilt Trip

Stardust

Clerks

Skyfall

Hansel and Gretel

Flyboys

Goonies

Temptation

Flight

Miss Congeniality

Spaceballs

Enigma

Eden Log

All Good Things

House

Grimm

Beetlejuice

Caddyshack

Star Trek: Wrath of Khan

Popular TV shows

The Americans

The After

Vikings

Downton Abbey

Orphan Black
Hannibal
Veronica Mars

HBO
Sapranos
True Blood
The Wire
Boardwalk Empire
Band of Brothers
Deadwood
Eastbound & Down
Six feet under
Treme
Enlightened
Oz

Original Content
Amazon premiered the comedies Alpha House and Betas.

Sling TV
https://www.sling.com/

7 days free.

Available on Amazon Fire TV, Roku, Mac, PC, Android, iOS

They offer online and device packages streaming to multiple devices.

The basic package is $20 per month. You get live TV and sports, plus movies, and news. With ESPN, ESPN2, TNT, TBS, AMC, A&E, IFC, Travel Channel, Maker, CNN, Adult Swim, Disney Channel, Food Network, History, H2, Lifetime, HGTV, Cartoon Network, El Rey, and Galavision.

For $15 more, you can get HBO.

They also have $5 packages for Sports (ESPN & SEC), Kids (Disney), Movies (EPIX & Sundance), News (International news), and Lifestyle (home and living channels)

Playstation Vue
http://www.playstationnetwork.com/vue/home/

$49.99 Basic, $59.99 Core, $69.99 Elite

Sony jumped into the game using their own console system to stream their own Pay TV channel packages to some users. Now available in Chicago, New York City and Philadelphia from PlayStation®Store on your PS4™ and PS3™. Be on the lookout for additional cities and availability as time progresses.

Access Price: **$49.99**	Core Price: **$59.99**	Elite Price: **$69.99**
Viewers can start enjoying more than 50 of the most popular channels of live TV, movies and sports with PlayStation Vue's basic starter package.	Core includes all Access package channels, plus the following local sports channels for select markets.	Elite provides the ultimate PlayStation Vue experience, and includes all channels from Access and Core packages, plus the following lifestyle, music and family channels.
Broadcast: CBS, CBS Plus, Cozi TV, Exitos, FOX, MyNetwork, NBC, Telemundo. **Network:** Animal Planet, BET, Bravo, Cartoon/Adult Swim, CBS, CMT, CNBC, CNN, Comedy Central, Destination America, Discovery Channel, Discovery Family, DIY, E!, Esquire, Food Network, Fox Business, Fox News Networks, FOX Sports 1, FOX Sports 2, FX, FXX, HGTV, HLN, Investigation Discovery, MSNBC, MTV, MTV2, Nat Geo, NBC Sports Network, Nick Jr., Nickelodeon, Nicktoons, OWN, Oxygen, Science, Spike, Syfy, TBS, TLC, TNT, TruTV, Travel Channel, TV Land, USA Network, VH1. AMC will be available in April.	**Network:** BTN, Golf Channel, TCM. **New York Only:** YES Network. **Philadelphia Only:** Comcast SportsNet Philadelphia. **Chicago Only:** Comcast SportsNet Chicago.	**Network:** American Heroes, BET Gospel, Boomerang, Centric, Chiller, Cloo, CMT Pure Country, CNBC World, Cooking Channel, Discovery Fit & Health, FOX College Sports Atlantic, FOX College Sports Central, FOX College Sports Pacific, FXM, LOGO, MTV Hits, MTV Jams, mtvU, Nat Geo Wild, PALLADIA, Sprout, TeenNick, Universal, Velocity, VH1 Classic, VH1 Soul.

Redbox Instant (EPIX) - Shutdown Oct 10, 2014

See message of shutdown http://about.redboxinstant.com/news/

Redbox Kiosk available titles. http://www.redbox.com/ also has links to iOS and Android apps

HBO Go

Because of this, I have added HBO to the app comparison chart replacing Redbox Instant subscription. HBO Go is supported by many devices.

HBO NOW

https://order.hbonow.com/

The new streaming stand alone service so far only available through apple iPhone and Tablet, and Optimum online service. Watch for future availability. Free first week. $14.99 each month. Coincided with the new season of Game of Thrones.

Crackle (Sony)

Crackle is backed by Sony Pictures Entertainment

Original Programs: Shows include "Bannen Way" "Comedians in Cars Getting Coffee" "Chosen", "Cleaners", "Trenches", "Unknown", "Sports Jeopardy", "Sequestered"

Original full feature film; "Extraction"

Movie and TV Library

Crackle features many Columbia Pictures and Screen Gems titles including Men In Black, Pineapple Express, District-9, Step Brothers, Talladega Nights, Blankman, Fletch, Conan.

Crackle features Sony distributed television series like "Damages", "Rescue Me", "The Shield" and "Seinfeld".

Crackle's content refreshes monthly with titles being added and taken down.

Content Partners
Aniplex
FOX Digital
Funimation
Lionsgate
MGM
Red Bull
SnagFilms
TOEI

Genres

Crackle features programming in the following key genres: Action, Comedy, Crime, Drama, Horror and Sci-fi.

In January 2012, Crackle added Animex anime channel to its content, available to viewers in the US and Canada.

Top Titles

Movies

Resident Evil: Afterlife

Sol

Attack the Block

The Roommate

Under Suspicion

Talladega Nights: The Ballad of Ricky Bobby

Slackers

Edison Force

Booty Call

Run Lola Run

Insanitarium

April Fool's Day

Candyman

American Psycho

Zonad

Kaena: The Prophecy

Godzilla 2000

TVShows

Jeopardy! Flashback

Queen's Blade

Marvel Anime: X-Men

The Jackie Chan Adventures

Barney Miller

Aim High

All In The Family

Bewitched

Blood+

The Border

Damages

Good Times

I Dream of Jeannie

Married With Children

News Radio

Rescue Me

YouTube Movies (Google: Merged with Google Video)
https://www.youtube.com/user/movies

You can buy movies, but here is a link to free movies offered on YouTube and a list of the main channels/studios.
Free YouTube
Movieshttp://www.youtube.com/user/movies/videos?sort=dd&shelf_id=12&view=26

Movies by
Starz Media (Anchor Bay) http://www.youtube.com/user/starzmedia
Maverick Movies http://www.youtube.com/user/maverickent
http://maverickmoviesonline.com/
Entertainment One Benelux http://www.youtube.com/user/eonebenelux
Gravitas Ventures http://www.youtube.com/user/GravitasVOD
Cinedigm http://www.youtube.com/user/NewVideoDigital
Abehorror http://www.youtube.com/user/Abehorror
I dream Productions http://www.youtube.com/user/IDreamProduction
Viewster http://www.youtube.com/user/ViewsterTV
FirstLookStudio (Millennium Entertainment) http://www.youtube.com/user/FirstLookStudios
Popcornflix http://www.youtube.com/user/ScreenMediaPictures
Cinama Nirvana http://www.youtube.com/user/CinemaNirvana
Docurama Films http://www.youtube.com/user/docurama
Drelbcom http://www.youtube.com/user/drelbcom
VISOCinema http://www.youtube.com/user/VISOCinema
MANGAEntertainment http://www.youtube.com/user/MANGAentertainment

Popcornflix (Screen Media)
http://www.popcornflix.com/

The site primarily consists of independent feature films, many of which come directly from Screen Media's library. No subscription is required rather it is ad based.

Top Movies
Dead Tone
The Cry
The Toxic Avenger
Class of Nuke em High
Easy
Two Brothers and a Bride

2 Days
Descent
30 Years to Life
Dream Warrior

TV Series
Monster Garage
Cheaters

National Geographic Channel
Life Below Zero
Icons of Power
Tools of the Trade

PlutoTV
http://pluto.tv/#!

A recent start up which takes YouTube content and turns it into more of TV like continuous content, curated by people. There are 100+ channels covering music, news & info, lifestyle, education, sports, comedy, tech, entertainment, art & culture, and kids. Although not new content, it serves as an intermediary and organizes the YouTube content into continuous content streams. It transforms some YouTube content into user friendly streams.

This is fun and works rather well.
Available PC, Mac, Amazon Fire TV, Chromecast, and Apple TV (as of 11/14/2014)

TubiTV
http://tubitv.com

Tubi TV content partners include Starz Digital Media, Cinedigm, Shine International, Jim Henson Co., Hasbro Studios, Film Movement, ITV, Endemol, Zodiak Rights, DRG, All3Media, Kino Lorber, Korean TV network MBC and Korean studio CJ Entertainment. In addition, Tubi TV has lined up several digital content partners, which include Newslook, AP, Reuters, anime distributor Funimation, Havoc Television, ACC Digital Network, Viki, Anyclip.com and Wochit.

It has some unique content not found elsewhere helping it supplement other services nicely.

Available Roku, Amazon Fire TV, Android, iOS

Snag Films
http://www.snagfilms.com/

Documentary and Independent films. National Geographic and PBS also places their produced films here.

Filmon
http://www.filmon.com/

Has live and VOD free content, with occasional commercial interruptions.

Live UK channels, horror, classic tv, some news channels, music, entertainment, gaming, lifestyle, sports, tech

With apps for Roku, PC, Android, Windows mobile, and iOS among others.

VUDU (Walmart)
VUDU is a video on demand (VOD) video delivery company owned by Walmart.

15,000 total titles in the VUDU catalog, including both movies and television shows. VUDU has established content licensing contracts with all major movie studios as well as over 50 smaller and independent studios.

They offer a 5 free video HD streaming movies from select titles.

VUDU can stream videogame consoles, many Blu-ray players with build in smart functions, iOS mobile and Android devices. Roku, Neo, and many streaming media devices.

New Rentals are $3.99 to $5.99 depending on definition.

New Purchases are $14.99 to $19.99 depending on definition.

TV series can be $1.99 to 2.99 per episode.

Target Ticket
Similar to Walmart's VUDU service they have similar prices as well. They offer 10 free video HD streaming video downloads at sign up from a pre-selected pool.

M-GO (Dreamworks and Technicolor)
Offers 2 free movies, after that pay-as-you go.

Partner of Dreamworks and Technicolor.

New Rentals are $3.99 to $4.99 depending on definition.

New Purchases are $14.99 to $19.99 depending on definition.

TV series are $1.99 per episode.

Sony Entertainment Network
Similar to other download movie services in price.

Fandor
Film festival type films. Independent, classic, foreign, and documentary. Available on many devices. Android, iOS, Kindle, Roku, PC & Mac

$90 annually or $10/month

ErosNow
http://erosnow.com/

Hindi movies. Some are free. Some require subscription.

DramaFever
http://www.dramafever.com/

Watch Korean dramas, TV shows, and movies free.

Return to Table of Contents

Sports Packages

Make sure to check whether or not your device will be able to use these sports package services.

MLB

http://mlb.mlb.com/mlb/subscriptions/

MLB.TV Premium $129/year, $24.99/month. Desktop & laptop as well as mobile and streaming devices. You get home _and_ away announcer audio feeds, mobile support, and At Bat 14 app for mobile included.

MLB.TV Basic $109.99/year, $19.99/month desktop or laptop only.

MLS

http://www.mlssoccer.com/mls-live

2014 prices - MLS $74.99/season Roku, Apple TV, iPad, iPhone, Android, & Panasonic Devices

NASCAR

http://www.nascar.com/en_us/ajax/static/raceview-product-page.html

2014 prices -

NASCAR Raceview Premium $59.95/season, $9.95/month. PC only. This is 3D virtual models rather than video feed. Statistics

NASCAR Raceview Audio $19.95/season, $4.95/month. PC only Broadcast and Driver audio.

NASCAR Raceview Mobile Premium $39.99/season, 4.99/month. 3D virtual race & audio of drivers and broadcast.

WNBA

www.**wnba**.com/**Live**Access

WNBA Live Access 2014 $9.99/ Season

NFL (Computers and Windows, Android, iOS Tablets)

http://www.nfl.com/gs/GameAccess/index.jsp

NFL Preseason $19.99. Live games except blacked out games, which you can watch 24 hours after the game completes.

NFL Game Rewind Season Plus $59.99/season or 4 payments of $17.99/month. Replays of every NFL Game, Playoffs, and Super Bowl (not live).

NFL Game Rewind Season $29.99. Replay of every NFL game during regular season (not live).

NFL Game Rewind Follow Your Team $24.99/season. Replay of every game of your favorite team during the regular season (not live).

https://audiopass.nfl.com/nflap/secure/packages?ttv=0

NFL Audio Pass Season Plus $19.99/season. Listen to every game live including Playoffs and Super Bowl. Listen to archived games.

NFL Audio Pass Season $14.99/season. Listen to every regular season game live. Listen to archived games.

NFL Audio Pass Follow Your Team $9.99/season. Listen to your favorite team's games live. Listen to archived games.

NHL
www.**nhl**.com/gcl/

NHL Gamecenter Live 2014-15 $99.95/season, or 5 payments of $19.99/month.

NBA
http://www.nba.com/leaguepass/

NBA League Pass - TV, internet, mobile $199/Season; Digital only- Internet and mobile $149 (although it's half off halfway through this season)

UFC
http://www.ufc.tv/page/fightpass

UFC Fightpass - $9.99/month, 6m commitment $8.99/month, 12m commitment $7.99/m. Live Fight Nights, UFC Unleashed, Best of Pride, The Ultimate Fighter, originals

WWE Network
http://www.wwe.com/wwenetwork

$9.99/month. Includes live all broadcasts streaming, original content, archives, & live pay-per-views.

Popular Free Sports Online and Apps
NFL Now & NFL Mobile- NFL Network clips, scores, and news   http://now.nfl.com/

Fox Sports - http://www.foxsports.com/

Watch ESPN - http://espn.go.com/watchespn/

Hunting, Outdoors, and Fishing - http://www.carbontv.com/

HuntIt.TV http://www.huntit.tv/ hunting, fishing

Return to Table of Contents

NETWORK TELEVISION (Online Website Links and Resources)

Commercial Television

*ABC   http://abc.go.com/   watch episodes

ABCNEWS   http://abcnews.go.com/   clips and news

*CBS   http://www.cbs.com/   watch episodes

CBSNEWS   http://www.cbsnews.com/   clips and news

*CW   http://www.cwtv.com/   watch episodes

*FOX   http://www.fox.com/   watch episodes

FOXNEWS (cable station)   http://www.foxnews.com/ clips and news

*NBC   http://www.nbc.com/   watch episodes

NBCNEWS   http://www.nbcnews.com/   news and clips

Public Telvision

*PBS   http://www.pbs.org/   full free episodes, news

CREATE   http://www.createtv.com/   schedule, episode info, clips, a few episodes

WILDTV   http://www.pbs.org/wnet/wildtv/

**MHZ   http://www.mhznetworks.org/   free full episodes, watch MHZ Worldview free live video streaming

*WORLD   http://worldchannel.org/ free full episodes mostly documentaries

Childrens

PBJ   http://watchpbj.com/ schedule

*PBS Kids http://pbskids.org/ schedule, clips, episodes

QUBO   http://www.qubo.com/home clips, games

*BOUNCETV   http://www.bouncetv.com/ Some full episodes of original content, schedule

IONTV   http://iontelevision.com/ clips, schedule

MYTV   http://www.mynetworktv.com/   clips

Movies

THIS TV http://thistv.com/ schedule

GET TV   http://get.tv/ schedule

MOVIES! http://moviestvnetwork.com/ schedule

ESCAPE TV http://www.escapetv.com/ schedule

Music

**ZUUS COUNTRY   http://www.zuus.com/ watch free live TV stream

Living

IONLIFE   http://www.ionlife.com/ clips, schedule

*LIVING WELL   http://livewellnetwork.com/ watch many free episodes

MYFAMILY TV   http://www.myfamilytv.tv/ schedule

Tuff TV   http://www.tufftv.com/ full episodes of Auto Wars original programming

Classic TV

RETRO TV   http://www.myretrotv.com/ schedule

*ME TV   http://metvnetwork.com/ some free full episodes, schedule

ANTENNA TV   http://antennatv.tv/ schedule

COZI TV   http://www.cozitv.com/ schedule

Religious

**Atheist TV http://atheists.org/AtheistTV free live TV

**BUDDHIST (Buddhism)   http://www.thebuddhist.tv/   free live TV and radio stream

**BYUTV (Mormon)   https://www.byutv.org/ clips, episodes, free live TV

**CBN (Christian)   http://www.cbn.com/   clips, episodes, free live TV

**GODTV (Christian)   http://www.god.tv/   episodes, free live TV

**EWTN (Catholic)   http://www.ewtn.com/   news, information, live TV and radio stream

**ISLAM CHANNEL   http://www.islamchannel.tv/   news, information, live TV

**JLTV (Jewish Life)   http://www.jltv.tv/ show info, free live TV

**JN1 (Jewish News)   http://jn1.tv/ news, free live TV

**SHALOMTV (Jewish)   http://www.shalomtv.com/

*means full episodes are available

** means live streaming is available

Return to Table of Contents

Paid TV channels (Online Website Links and Resources)

*Has some free unlocked episodes.

**Has live streaming unlocked.

*A&E   http://www.aetv.com/   some full episodes, some locked, clips

ABCFamily   http://abcfamily.go.com/   watch live and episodes with TV provider permission

AHCTV   http://www.ahctv.com/   clips, schedule

AMC   http://www.amctv.com/   full episodes with TV provider permission, extras clips free

Al JEZEERA http://america.aljazeera.com/ clips

*Animal Planet   http://www.animalplanet.com/   some clips, newest finding bigfoot

Anime Network   http://www.theanimenetwork.com/ $6.95/month online

AXS TV http://www.axs.tv/ schedule

BBC America   http://www.bbcamerica.com/   some clips

BET   http://www.bet.com/   show clips and exclusives

*BIO   http://www.biography.com/   some full episodes, full biographies, and mini biographies

**BLOOMBERG   http://www.bloomberg.com/   free live stream USA, Europe, Asia, Event

*BRAVO   http://www.bravotv.com/   clips and some full episodes

**BYUTV (Mormon)   https://www.byutv.org/   clips, episodes, free live TV

**CBN (Christian)   http://www.cbn.com/   clips, episodes, free live TV

ChillerTV   http://www.chillertv.com/   clips, schedule

CLOO   http://www.cloo.com/   clips

*CMTV   http://www.cmt.com/   clips and some full episodes

*CNBC   http://www.cnbc.com/   many full episodes, watch live with TV provider permission

*CNN   http://www.cnn.com/   many clips by show, watch live with TV provider permission

*COMEDY CENTRAL   http://www.thecomedynetwork.ca/   full episodes of main shows

*COOKING   http://www.cookingchanneltv.com/home.html   recipes, watch some episodes

**CSPAN   http://www.c-span.org/   free Live CSPAN, CSPAN2, CSPAN3, CSPAN Radio, also clips

DESTINATION AMERICA   http://www.destinationamerica.com/   clips

*DISCOVERY   http://www.discovery.com/   many Mythbuster and other full episodes, clips

DISCOVERY FIT & HEALTH   http://www.discoveryfitandhealth.com/   clips

*DISNEY   http://disneychannel.disney.com/   some free full episodes, watch live with TV permission

E!   http://www.eonline.com/   clips and news

**ESPN   http://espn.go.com/   clips and news, Watch live ESPN3, Watch other ESPN with TV provider permission

ESQUIRE   http://tv.esquire.com/   watch live and episodes with TV provider permission

*FOX NEWS   http://www.foxnews.com/   clips and news, watch Fox News and Fox Business with TV provider permission

Fox Sports   http://msn.foxsports.com/   clips from shows

*FOOD NETWORK   http://www.foodnetwork.com/   many free full episodes

FUSE   http://www.fuse.tv/   clips

FUSION http://fusion.net/ schedule

*FX   http://www.fxnetworks.com/   clips and some free full episodes

FXX   http://www.fxx.com/   schedule

FXM   http://www.fxnetworks.com/fxm   schedule

G4   http://www.g4tv.com/ clips and news

*Golf Channel   http://www.golfchannel.com/   free full episodes, live with TV provider permission

*GSN (Game Show)   http://gsntv.com/   one of each full free episodes, clips

HALLMARK   http://www.hallmarkchannel.com/   clips

HALLMARK MOVIE   http://www.hallmarkmoviechannel.com/   schedule

*HGTV   http://www.hgtv.com/   many free episodes

*HISTORY   http://www.history.com/   many free full episodes

*H2 (History2)   http://www.history.com/shows/h2   many free full episodes

*HLN   http://www.hlntv.com/   clips and news

**HSN, HSN2   http://www.hsn.com/   all the shopping items, free live TV stream

*IDTV   http://www.investigationdiscovery.com/ full Episodes, clips

*IFC (International Film Channel)   http://www.ifc.com/   they have a free stream room

*LIFETIME   http://www.mylifetime.com/   some free full episodes, some with TV provider permission, some full lifetime movies

*LMN   http://www.mylifetime.com/movies/lifetime-movie-network   a few full episodes

*MLB TV & AUDIO   http://m.mlb.com/network   scores, news, registered user audio, subscriber video stream

*MSNBC   http://www.msnbc.com/   show clips, news, live with TV provider permission

*MTV   http://www.mtv.com/   news, clips, some full episodes, popular music videos

National Geographic   http://channel.nationalgeographic.com/   video clips, locked episodes

National Geographic Wild   http://channel.nationalgeographic.com/wild/   video clips

NBA TV   http://www.nba.com/nbatv/   clips, news, scores, league pass

NFL   http://www.nfl.com/nflnetwork   clips, news, scores, audio and video subscriptions

NHL   http://www.nhl.com/ice/eventhome.htm   clips, news, scores, subscriptions, archives

*NICKELODEON   http://www.nick.com/   many free full videos, games

*NICK JR   http://www.nickjr.com/   activities, games, video clips

*TOONS   http://nicktoons.nick.com/   many free full episodes, clips

OUTDOOR   http://outdoorchannel.com/   clips, info

OXYGEN   http://homepage.oxygen.com/   some clips, episodes with provider permission

OWN   https://www.oprah.com/own   clips

PALLADIA   http://www.palladia.tv/   just schedule

PIVOT TV http://www.pivot.tv/ clips, schedule

*POP TV http://poptv.com/ episodes, schedule

**QVC   http://www.qvc.com/   all the shopping items, free live TV stream

REELZ   http://www.reelz.com/   clips

REVOLT TV http://revolt.tv/ clips

**RT News (Russia Today)   http://rt.com/   news, free live TV http://rt.com/on-air/

SCIENCE   http://www.sciencechannel.com/   clips

**SHOPHQ   http://www.shophq.com/   all the shopping items, free live TV streams

*SPIKETV   http://www.spike.com/   many free full episodes

*SMITHSONIAN   http://www.smithsonianchannel.com/   some full episodes

SPEED   http://msn.foxsports.com/speed   clips and news

SUNDANCE   http://www.sundance.tv/   watch with TV provider permission

*SYFY   http://www.syfy.com/   many free full episodes

TBS   http://www.tbs.com/   many full episodes and live with TV provider permission, Ground Floor without permission

TCM   http://www.tcm.com/   watch on demand and live with TV provider permission only

TENNIS   http://www.tennischannel.com/   clips, watch live with TV provider permission

*The Weather Channel   http://www.weather.com/   personal weather info, clips

*TLC   http://www.tlc.com/   a few full original series episode, clips

TNT   http://www.tntdrama.com/   live and episodes with TV provider permission

TruTV   http://www.trutv.com/   clips, episodes with TV provider permission

*CARTOON Network   http://www.cartoonnetwork.com/   some clip, 1 free episode no permission for 10 cartoon series, others require with TV Permission

*TRAVEL   http://www.travelchannel.com/   some free episodes and clips

TVGUIDE NETWORK   http://www.tvgn.tv/   some free full episodes, clips

*TVLAND   http://www.tvland.com/   many free full episodes

*TVONE   http://tvone.tv/   some full episodes

UPTV (formerly Gospel Music Channel)   http://www.uptv.com/music   clips and video

USA   http://www.usanetwork.com/   some free full episodes, most episodes with TV provider permission

VELOCITY   http://www.velocity.com/   clips

*VH1   http://www.vh1.com/   many free full episodes

WWE   http://www.wwe.com/   clips, news, WWE Network with subscription

HBO   http://www.hbo.com/   clips, HBO subscription coming in 2015

CINAMAX   http://www.cinemax.com/   clips

SHOWTIME   http://www.sho.com/   clips

The Movie Channel (owned by Showtime)   http://www.sho.com/site/tmc/videos/home.do

ENCORE   http://www.encoretv.com/   trailers and clips

*STARZ   http://www.starz.com/   free full <u>first</u> episode only of each series

Return to Table of Contents

NOTES